MY PERSPECTIVE...

Insights on
Life, Learning, and Decision-Making

Tony Pray, PMP

The Insight Series Book 1

TotalRecall Publications, Inc.
1103 Middlecreek
Friendswood, Texas 77546
281-992-3131 281-482-5390 Fax
www.totalrecallpress.com

AI Illustrations: "Illustrations by Dall-E"
Book Cover Design: by Tony Pray
The Insight Series™

ISBN: 978-1-64883-404-2
UPC: 6-43977-44042-6

FIRST EDITION
1   2   3   4   5   6   7   8   9   10

To Donna,

This book is dedicated to my loving wife and life partner of the past 39 years - and the next 50 or so, Donna. Together, we are an exceptional being.

You gave me my reason to be the best man I could be. You've challenged me, supported me, built me up, and put up with my personal brand of craziness. You've been my rock, my greatest cheerleader, and my partner in every sense of the word. We've faced life's challenges together for nearly four decades—lifting each other when we've fallen and laughing through the tears. You make me smile every day. With all my love and gratitude for the life we've built and the future we continue to create, this book is for you.

Table of Contents

Authors Bio

Tony was born when free-range kids came home when the street lights turned on, raised during the racial tensions and antiwar demonstrations of the sixties, and forged into the brilliant and kind person I am today (per my wife) by 73 years (and counting) in the crucible of ups and downs of life. I have been married for 39 years (so far) to my wonderful wife, Donna. I've had dozens of jobs ranging from Sergeant in the Marine Corps to motorcycle salesman, lineman, truck driver, electrician, railroad worker, avionics crew chief, telecommunications engineer, call center manager, computer support manager, hobby farm blacksmith, energy auditor, building inspector, and for about 30 years, senior IT Project Manager.

With my wife, Donna, I raised our blended family of 4 children, who range from 55 to 43 years old. They, in turn, have given us 5 grandchildren, and we just welcomed our first great-grandchild. I'm a committed lifelong learner, martial arts instructor, guitarist, and writer. My previous book, **A Concise Guide to Better Decisions**, is still in print. I'm retired, although I'm still busy writing my blog, "**My Perspective,**" on SubStack, target shooting, and reading voraciously. I develop AI prompts and maintain an AI prompt library for educators.al.

Preface:

Throughout my career as a Senior IT Project Manager, I often found that the biggest hurdle to thriving with technology wasn't the technology itself—it was finding the entry points to really understand it. Complex ideas, loaded with jargon, created barriers rather than bridges. My hope in writing this book is to empower readers to thrive using insights that are accessible, practical, and even a little bit fun.

While technology underpins much of our modern world, it's the human element that truly drives progress. That's why these pages blend real-life stories, practical strategies, and a touch of humor—so we can explore and apply new ideas without the headaches.

Expect clear explanations, actionable tips, and fresh perspectives designed to help you thrive in today's ever-evolving just plain crazy landscape.

Thank you for taking the time to join me. I hope these insights spark curiosity, instill confidence, and give you food for deep thought. Happy reading—and welcome to a new way of thinking about what's possible.

11-27-23
The magic of simple rules of thumb

Life in EASY mode

The magic of simple rules of thumb

Straightforward rules of thumb are my secret sauce because:

1. They are easy to understand and communicate.

2. They reliably increase your chances of success.

3. They allow for speed, agility, and grabbing opportunities while reducing risks and regrets.

4. **They only apply when the stakes are low enough**. After considering the Consequences, Risks, and Costs of a decision, **the stakes won't ruin me even if things go completely wrong.**

   Note: Simple rules of thumb are not appropriate for every decision. Some decisions carry high risk, impact, or cost. These are serious decisions. Every "Serious" decision deserves your full attention.

   Since we are examining rules of thumb, here's **my rule of thumb for triaging decisions:**

1. If a decision has **any** of the following characteristics:

2. High **impacts** on myself, or others

3. High **risks** (financial or other including reputational risks)

4. High **costs** or other serious **consequences**

5. **Permanent** or long-lasting effects upstream or down

6. **Irreversible** or impossible to try out in a small pilot

It is an important decision. — Take it through a complete decision process.

The 80% Rule

If a decision does not have any of these characteristics, I consider it an **everyday decision and I am free to use good rules of thumb** to speed up my decisions.

Most decisions are simple and low-risk and do not need a formal decision process. Well-chosen rules of thumb will help position you properly to avoid or reduce risks and take advantage of opportunities. This is the zone where 80% to 90% of your decisions will live.

What if you could make those remaining decisions just a little better? Doesn't seem like much at first glance, but small decisions accumulate over time to create real long-term changes in your life. **Each decision you make positions you differently for the next thing coming along**. A small decision to have dessert after every meal doesn't mean much at the time but a few months later…

Small changes affect what your next starting point is going to be. Over time, they create a compounding effect that works exactly like compounding interest. Small changes compound to produce large results.

Small decisions directly affect your positioning. My friend Robert (name changed) had just had a new baby when his job was unexpectedly eliminated. He had the stress of looking for a new job and providing for his family while going on very little sleep. One day, he had a flat. Not a big deal, he just changed the tire and went on his way. However, he didn't have the money right away to get the old tire fixed. Life stayed busy and a few months later, he was driving to a seminar when he had another flat. This time he had no spare. A tiny decision that eventually cost him the price of a tow truck and a missed seminar. **Small decisions have an outsized effect on your life because they affect your starting point.**

Rule of thumb #1 — Before I decide anything — **Make sure I'm fit to decide** using the H.A.L.T. rule: If I'm HUNGRY, ANGRY, LONELY, or TIRED — or if I'm overly excited or under any undue pressure I stop and regroup. I make it a personal rule to "sleep on it" in these situations.

Rule #2 — **Ask "What information will I need to make a good decision?** I overlook important factors if I don't think through what the key information is and what is irrelevant.

Rule #3 — **Get a clear understanding of the problem**. Make sure I'm solving the real problem and not just a symptom.

For the least consequential decisions, such as what to have for dinner, I try to **adopt good habits and make them into already-decided things. This pre-commitment** makes most choices easier. For example, every Tuesday evening after class, dinner is scrambled eggs and toast. (Because I'm usually exhausted after 3 hours of Aikido practice and teaching and because it has to be a light meal that's high in protein, but mostly because it's tasty and easy.)

Another example is to decide in advance you'll only eat dessert on Sundays. **Make the decision when you're at your best and make it a personal rule**. Other people don't seem to press as much when I say "I make it a rule to [insert personal rule here] ".

"Does it pass the sniff test?" To me, this means it lines up with the things I value, doesn't put my reputation or my opinion of myself at risk, and has no parts that are unethical or immoral. **Every** decision must meet my own standards, not just the most important ones.

When any part of a decision doesn't pass my sniff test, I make it a personal rule to reject it altogether or adjust the decision accordingly.

When you're trying to decide between competing alternatives, take a minute to **decide what's important and pick only from those that meet all of your minimum standards** in the criteria you consider important. Anything you pick will at least be adequate.

Always make certain to choose from at least 3 options. Decisions like "Should we fire him or not?" are binary. Considering just one extra option (perhaps neither of the above is the best choice) can prevent you from coming to premature solutions.

To make a decision that's as unbiased as possible, **imagine that you don't know where you'll be born**, what race, sex, abilities and handicaps, or anything else you'll have. Would this be a fair decision for you?

Does this decision follow the platinum rule? Treat others as they would like to be treated. Does this decision maximize overall well-being and minimize suffering?

Many rules of thumb are useful only in narrow ways and some of them contradict others. For example, "Penny wise, pound foolish" contradicts "A penny saved is a penny earned." Select your own rules carefully.

I curate my own rules of thumb based on these criteria:

1. Does this rule reliably improve my chances of a good decision? Is it better than a coin flip?

2. Can this rule be applied broadly?

3. Does this rule encourage me to think of better alternatives?

4. Does it position me better for the future? Good positions create options, while bad positions reduce them. You don't have to be an expert decision-maker to get better results, you only need to put yourself in a good position. Anyone looks like a genius when all the options are good.

5. Is it a rule that helps or minimizes harm to others? Is it prosocial?

6. Is it a proactive rule? Can you act on it to improve your position?

7. Does this rule fit into a systematic approach to making good decisions in general, even for trivial things?

8. Does it increase my chances of learning something while decreasing risk? Dipping a toe in the water is an example.

9. Does it help you find possible win-win solutions?

   (BTW: I'm currently experimenting with prompts to teach an A.I. how to help me curate great rules of thumb by scoring any given rule based on the criteria above. It's given me some very unexpected results so far.)

   That's My Perspective...

12-04-24
On Being A Real Pro...

What it takes to become the true pro
you really want to be.

"Illustrations by Dall-E."

Executive Summary

Everyone wants to be perceived as a professional in their
work. From here on, I'll refer to a true professional as a
Pro (capitalization intentional). Here are the things to pay
attention to:

Expected Results

- Better perception by others

- Better reputation

- Trust builds more quickly

- Better results

- Respect from others

Background

After leaving the Marine Corps, I worked at a variety of technical jobs including avionics crew chief and lineman. I became a telecommunications technician and then manager. Over time, I went up the ranks to become a SR. IT Project Manager with responsibility for many multimillion-dollar projects all over the globe.

Communicating directly with the C-suite executives, our Fortune 100 customers and just as often working with the techs and engineers doing the work on the ground. I had a successful career doing work I loved.

I worked for some horrible bosses, some good ones and some excellent ones. I was a pretty horrible boss at one time. I learned slowly due to my lack of understanding about what a true pro really is.

When I found myself in a setting where I was the low man on the totem pole in a civilian job, I finally started to learn what I needed to do about it...

Key lessons...

Starting parameters

One approach to finding out what a real pro looks like is to understand what a really poor example looks like. I've worked for and with some really bad examples and they have some characteristics in common.

I've seen them at all levels in dozens of companies. **The first guarantee a lack of professionalism is arrogance.** It is a self-centered and self-serving disdain for other people. It's often disguised toward superiors in the hierarchy but far too obvious to the people at or below that level.

Strangely the arrogance of some is misinterpreted as strength. It's the person who takes all the credit for the work his people did or even for the work of others.

So, in my definition of a "Pro," **humility is a top requirement**, a willingness to share credit and build other people up. A humble person treats others with respect, always. This is the person who understands it's a mark of respect to others to show up on time for the meetings, to be present and not focused on the phone or the last problem when dealing with a peer. It's an ability to see things from the other person's viewpoint.

This arrogance should be distinguished from pride or perfectionism, in the Marine Corps. From my limited exposure over a 3-year enlistment, I never met an NCO or officer I'd consider to be an example of arrogance even though the appearance is there.

Those people form the basis of my definition of real pros. They are dedicated and committed to their work in a way you'll find very hard to emulate in a civilian job.

Appearances

Appearances matter more than you think and a Pro knows this. It starts with personal grooming. You can't project competence and ability if you can't even maintain your own appearance. There's a reason most corporate CEOs today still wear business suits. Even as a very young phone technician, coming to the job clean and well-kept, made it easier to be taken seriously by my clients.

Keeping your tools in great shape, your desk clean, your surroundings organized make you more effective and also make you APPEAR to be more effective.

The appearance of calm and being in control (The reality is even better if you can do it.) does more to help morale in difficult times than anything else. It builds confidence and allows people to be at their best when the worst happens. Master yourself to be a Pro.

Show Up Prepared

You can't appear at all, unless of course you **SHOW UP**. Showing up, doing the work, being on time, delivering on time with craftsmanship and care. Doing that Every Time. These are the hallmarks of a Pro.

If you show up and don't have a single clue about what to do, you have just lost any chance you may have had to call yourself a Pro.

I was once tested by my Executive Director when I had just been a Project Manager for him for a short while. He was taking a vacation and put me in complete charge while he was gone for 2 weeks. I was immensely flattered and ready to set the world on fire until I attended the very first meeting on his calendar. It involved an upcoming merger (we went through 9 of them while I was there) and technical joining of 2 telecom companies. I could barely catch the conversation as new acronyms flew by, technical issues I'd never heard of were discussed and financing decisions were presented.

What really popped my bubble was that all of this information was being presented to ME! I was in charge and I was responsible for making the decisions. I had not understood any of them well enough to even comment.

I did the only thing I could reasonably have been expected to do. I asked the team making the presentation for their recommendations. I was certain I'd be out of a job when the boss got back but I made the call they had recommended.

Unknown to me, the boss had set all of this up as an object lesson for his protege' and it did serve the purpose of damping my raging ambition a little and acquainting me with the reality of his job. It also gave him the chance to see how I operated when I was out of my depth. I got lucky in this case since he had explicitly instructed his team to make sure I wasn't allowed to do too much damage. I didn't know the guard rails were up though.

The boss had demonstrated a couple of the key things a Pro needs, **foresight and a willingness to groom the next generation.** The foresight he demonstrated was to put up guardrails which meant I could not fail or damage the company and still get the job of orienting me done.

I learned an important lesson about being prepared. From then on, whenever I took on a new job, I first learned my own job in detail (documenting it as I went so I could teach it later if needed). As soon as I got the basics mastered, I started to learn how to do the job my boss was doing and to understand his or her world. That has served me very well.

Direction

Whether you are responsible to set the direction or to implement the strategy that results or just to follow the direction you're given, you must always be working toward reaching those goals or moving in the proper direction.

My son described this very aptly when he described a directionless life as sort of a "turd floating in a toilet".

While the metaphor might not be very pleasant, the effect of drifting with whatever currents come along can be just as unpleasant. If no one knows what direction you're trying to go in, they can't help you. If you yourself don't know what direction you're taking, your destination will always be random and you'll end up floating among other directionless people.

Promises and Trust

They say that "**business moves at the speed of trust**". That's why a team that's been together for awhile that has had time to get to know and trust each other can always outdo a team that just came together. If you trust that I have your best interests at heart when I ask you to do some unpleasant task, you're much more likely to give it your best shot.

Your reputation is something that takes a long time to build and only a moment to destroy. Guard it extremely jealously. Never let your integrity come into question. As Caesar's wife was told, "avoid even the appearance of scandal". The trust you work hard to build will let you succeed where anyone else would fail.

Great Bosses

I've worked for some people I would call Great Bosses (Shout out to Steven Smith!). **The first requirement is that they be real Pros to begin with.** They each had a real connection with their teams and put the needs of the team above their own needs. However they always kept the needs of the company in mind. They also have the burden of being excellent communicators and coaches.

Summary

Here's what you'll need to be a Pro:

Be humble. Appearances matter, especially calmness. Show up, do the work, be on time, deliver on time with craftsmanship and care. <u>Do that every time</u>. Be consistently reliable. Know what direction you want to go in. Build trust and lift up the people you encounter. Master yourself first.

Thanks for reading today's post. I hope you've found it helpful and thought provoking.

That's My Perspective...

12-11-23
Thinking From First Principles

Start with the basic rules of the universe...

Photo by Tengyart on Unsplash

Thinking from first principles helps you find creative approaches to problems that don't ignore the laws of the universe but do help you sidestep false assumptions.

To think for myself from first principles, I worked through this reasoning chain:

- The components of any choice that you make are (a) the skills you bring to the selection, (b) random chance/luck, and (c) your starting position.

- I am a committed lifelong learner, so I'm doing most of what I can do about my skills.

- Nothing can be done about random chance other than **positioning** to avoid or reduce risk and increase the upside of any choices.

- I need to clearly define my values and priorities. Those will guide what positions I will work toward.

- Once I know my direction, what actions are needed to move toward those aspirations?

- What are the most important values to me? Here are some of my starting points:

1. Start with health. No health, no good options.
2. Spend less than I make.
3. Take care of family and friends.
4. Change my environment to make positive movement easier, and negative movement harder. (e.g. add friction for snacking, buddy up for exercise, automate savings)

   - Develop an "operating system" for myself to make myself as resilient and adaptable as possible. Pre-commit myself to most routine decisions while I'm at my best. Engrain good habits that move me toward my aspirations.

That's My Perspective...

12-18-23
A Personal Operating System...

Setting myself up for success.

Finding the best way to handle day-to-day living.

Creativity and energy are wasted reinventing the wheel. So, I chose to curate the best of what's already been developed for my starting point. (**Positioning is important**).

A set of principles to guide me wasn't a simple matter of going with my default W.A.S.P. background, a particular religion, or other philosophy. So, I put together a do-it-yourself moral compass and principles kit. That meant asking some questions…

My first questions to myself were "What should I use as my starting point?" Should it be religion? Which one(s)? Philosophy, Science, Psychology, "Common" sense, or some other approach altogether? What about **a synthesis of the best ideas from wherever they originate**? I'd have to determine how to evaluate ideas, but this seems reasonable. Let's dig in further…

Religion as a guide?

I am cynical about religions, though I still value the feelings of community and the high aspirations of most. Also, "Thou shalt not kill" seems like good advice. I freely take the best of what religions offer, realizing they are run by humans with all our faults.

What about philosophy?

Some serious exploration brought me to the concepts of Stoic philosophy. I had heard it was a philosophy for slaves from ancient Greece. I wanted nothing of being guided by slaves and believed the stereotype of a stoic as some sort of emotionless zombie.

However, once I took a closer look to see what value I could get from it. Almost every concept resonated with me. Here are a few examples:

1. You have no control over the random things that happen due to chance. You do have control over how you react to them.

2. Energy spent on things you have no control over is wasted. Use that energy to change your actions to adapt to reality as it exists, not as you would like it to be.

3. The four main virtues of Stoicism are wisdom, courage, justice, and temperance. You should always demonstrate virtue.

4. Prepare yourself to deal with sickness, death, hunger, poverty, and the loss of everything. We are all mortal, and we will all die. Be prepared for those things so you can remain effective and useful when they happen. Think about them to appreciate the life you have and make the best of it.

5. You are responsible for the results in your life. When obstacles appear, you must adapt. When luck goes badly, you must recover and press on. You control your reactions to chance. You control the direction of your life. If you are moving in the right direction, you will get closer to your goals.

What about science?

Hell Yes! Taking useful concepts from a variety of religions and ancient philosophy is a great starting point (Think of it as my initial positioning.), but there are no

guarantees that they will meet the needs of the future. Adopting a scientific approach and willingness to experiment and learn has led directly to the most success our species has ever seen. A great rule of thumb is to learn from the most successful. Science has the best track record by far.

I've spent years studying formal Decision Science and have published my first book *"A Concise Guide To Better Decisions"*. Decision science gives us the most reliable course for making difficult and important decisions, but it takes **far too much time and effort for routine choices** that could still benefit from an upgrade.

How about handling my day-to-day defaults and habits? Rules of thumb?

Having a guide for the big decisions in life is great, but we all face hundreds of routine choices that could be improved. If 80 to 90 percent of your choices are routine and you could improve the average outcome of those choices by just 10 percent, imagine how the trajectory of your life would improve over time.

"Avoiding stupidity is easier than seeking brilliance" - Charlie Munger

I've been collecting, evaluating, and curating the most effective rules of thumb I can find. The simple act of thinking about what rules of thumb apply best to a situation helps me slow down and engage the logical part of my mind. There are thousands of commonsense rules that experience has taught in the only school that truly educates us all… life.

- **Summary:** Building a personal "operating system" for me starts with basic concepts.
  - from religion — "Thou shalt not kill", "Thou shalt not steal", etc.
  - from philosophy — Pursue the Stoic virtues.
  - from science — The scientific method and Decision Science,
  - from mathematics — Probability and game theory.
  - from experience — Rules of thumb, heuristics, and commonsense.

In my next post, I'll begin a deeper dive into how I select the rules of thumb and some of the personal rules I commit myself to follow.

Until then a good rule of thumb to remember is "Treat others as they would want to be treated".

That's My Perspective...

12-25-23
Weird Science

Let's play with a shiny new toy for today's article. **I am using an A.I. to help as I create this piece.** I'm feeling competent to explore since I've had a few hours of experimenting. As I go, I'll document the steps I'm taking and the missteps I make. I expect it will be a strange ride, so buckle in!

First, I have a topic already selected for today. This article will be a dive into **selecting good rules of thumb** (or heuristics) to help with life. I'll use the free Bing Chat, which uses ChatGPT version 4 when it's in creative mode.

"Illustrations by Dall-E."

After opening the browser window with Bing, I go to the chat screen and select "creative" mode. I then enter: "Please create several examples of illustrations to describe A.I. working alongside a person. Try for a hyperrealistic and flashy effect." I got four options and liked this one best. The guy looks like he's holding a cigarette there, but I didn't want to risk messing up the rest of the picture to remove it.

Next, I want the **AI to generate a couple of lists** for me — stages of life and a list of professions. I'll use these in a little while to help search for rules of thumb relevant to a wide range of professions and differing life stages.

My first stab at getting a list of life stages starts with this prompt: *"Please create a list of stages a person usually goes through in the course of a long and eventful life."* and the results are not quite what I had in mind. Here's a sample:

1. **Infancy (0-2 years)**: A time of rapid physical and cognitive development, marked by milestones such as first words and steps.

2. **Early Childhood (3-5 years)**: Continued growth and the beginning of formal education, with the development of social skills and friendships.

3. **Middle Childhood (6-11 years)**: Further academic learning, honing of interests and hobbies, and understanding of complex social dynamics.

That's not what I had in mind, so I added some examples to the prompt: "Please create a list of stages a person usually goes through, such as child, teen, young adult, marriage, career, in the course of a long and eventful life." and the results are getting closer to what I want but still not quite right:

Childhood: A time of learning and discovery, where one begins to form their first memories and understand the world around them.

1. **Adolescence:** Marked by the onset of puberty, this stage involves significant physical, emotional, and social changes, as well as the search for personal identity.

2. **Young Adulthood:** Often associated with pursuing higher education, starting a career, and establishing independence from parents.

3. **Marriage:** For many, this stage involves committing to a life partner and starting a family, though it's not a universal experience.

A final tweak to the prompt yields what I was looking for: "Please create a list of stages a person usually goes through, such as child, teen, young adult, marriage, career, in the course of a long and eventful life. List only the stages, not definitions or explanations."

My list is returned: Here is a list of life stages:

- Child
- Teen
- Young Adult
- Marriage
- Career
- Parenthood
- Middle Age
- Retirement
- Senior Years

That's what I wanted for the first list. I copy it to my notes for later use. Notice that in this case, the prompts and responses took more work than simply typing out the list would have taken. Worthwhile though, because I learn as I go.

Next, I want a **list of professions** which I'll step through later in my search for rules of thumb that are specific to a given career. Here's the initial prompt: *"Please create a list of at least 30 professions in a wide variety of categories. Make sure to represent knowledge workers such as project managers and scientists, teachers, students, first responders, etc. Just the generic name for the professions will suffice."*

That prompt produced this list, which was much quicker than typing one out:

1.	*Project Manager*
2.	*Scientist*
3.	*Teacher*
4.	*Student*
5.	*Firefighter*
6.	*Police Officer*
7.	*Paramedic*
8.	*Nurse*
9.	*Doctor*
10.	*Engineer*
11.	*Software Developer*
12.	*Graphic Designer*
13.	*Architect*
14.	*Lawyer*
15.	*Accountant*
16.	*Chef*
17.	*Electrician*

18.	*Plumber*
19.	*Carpenter*
20.	*Mechanic*
21.	*Pilot*
22.	*Flight Attendant*
23.	*Librarian*
24.	*Social Worker*
25.	*Pharmacist*
26.	*Veterinarian*
27.	*Dentist*
28.	*Researcher*
29.	*Journalist*
30.	*Marketing Specialist*

Now that I have my starter lists in hand, I'll write a prompt to step through each list and search through available resources for rules of thumb or short, pithy, and memorable mental rules, tools, and models to guide decision-making (when the stakes are low enough that it won't ruin you) to improve your odds of a good outcome.

Here's the prompt I used: "Using the list of life stages and the list of professions from earlier in this thread, please use your best available resources to create a list of "rules of thumb", "mental models", mental tools and ideas that are helpful and may be particularly helpful within one or more of the listed life stages or professions. The "rules of thumb" should be concise, memorable and clearly express the rule. Step through each stage and profession and provide at least 3 rules each."

Jackpot! I have a great list of rules as my starting point for the next post. In it, I'll continue my AI journey as I create a tool to evaluate and score each rule of thumb based on the criteria I will create.

Until then, here are a few examples of the rules of thumb we have to work with so far:

Life Stages:

Child:

Curiosity is key: Encourage exploration and asking questions.

Play is learning: Use playtime as an opportunity for growth.

Routine builds security: Establish consistent routines for comfort.

Teen:

- **Peer relationships matter**: Foster healthy friendships and social skills.

- **Independence is earned**: Gradually increase responsibilities.

- **Resilience over perfection**: Emphasize effort and learning from mistakes.

Professions:

Project Manager:

- **Clear goals drive success**: Define and communicate project objectives.

- **Risk management is crucial**: Anticipate and mitigate potential issues.

- **Stakeholder engagement**: Keep all parties informed and involved.

"Illustrations by Dall-E."

Scientist:

- **Skepticism fuels discovery**: Question assumptions and seek evidence.

- **Collaboration accelerates progress**: Work with peers across disciplines.

- **Publish or perish**: Share findings to advance knowledge.

With about 90 or so rules of thumb as a starting point next time, I'll work through prompting to score and rank the rules.

That's My Perspective...

01-01-24
Deeper into AI

In which we continue feeling our way through an A.I.-assisted post...

Happy New Year 2024!!

I predict that this is the year when A.I. changes everything, whether we realize it or not. In that spirit, let's continue to explore using A.I. to help write this series.

Idea Generation:

In the last post, we used Bing's free chat tool to create an illustration and to make lists of life stages and professions to be used later. This exercise sparked a series of ideas, showcasing **one of A.I.'s most valuable applications: the ability to ignite creative thoughts for further exploration.**

Here's my next prompt: "Please create a list of "passages" a person might expect to go through during their life. Examples include illnesses, financial troubles, loss of friends and family, or positive changes such as weddings, births, graduation, etc."

...and a sampling of results:

- Birth of a sibling
- Starting school
- Making a best friend
- First love
- Graduating high school

- Leaving home
- Starting university or a job

Then, I edited my earlier prompt to incorporate these "passages" into this work and generate some more rules of thumb to work with. Here's a sampling of the first few:

Birth of a sibling:

- **Embrace change**: Welcome new family dynamics with an open heart.

- **Share and teach**: Use this opportunity to teach and learn from each other.

- **Patience is a virtue**: Adjusting to a new sibling takes time for everyone.

Starting school:

- **Lifelong learning:** Approach education as a continuous journey.

- **Friendship fosters growth:** Cultivate positive relationships.

- **Resilience overcomes obstacles:** Persist through challenges.

While **these rules are very generic and lack specific actionability,** they serve as a foundation. The next step is to refine them into a set of 'keepers' that meet our established criteria.

What criteria should I use? Now that I have a substantial list of rules, it's time to winnow them down to a useful set. So, here's the next prompt I'll try:

"Using the rules of thumb created from earlier in this thread, and generally accepted rules in society such as the golden rule, please use your best available resources to rewrite each rule based on the following criteria:

1. Each rule must be actionable and stated in a way that provides an example action step a person can take to implement it.

2. Each rule must increase the odds of a good outcome, in other words, it must deliver results better than random chance.

3. The "rules of thumb" should be concise, memorable, and clearly express the rule.

Step through each rule as it is rewritten and score the result on a scale from 1 to 10 on the following parameters:

1. General usefulness (can it be helpful in many domains?)

2. How much does it contribute to the thriving of the individual?

3. How much would it contribute to society as a whole?

Once the entire list of rules is created, provide a summary listing of scores and top ten rules."

The initial results were somewhat unexpected:

Curiosity is key (Childhood)

- **Actionable Rule**: Ask "why" five times when encountering new information to deepen understanding.

- **Scores**: General usefulness: 8, Individual thriving: 9, Societal contribution: 7

- **Play is learning** (Early Childhood)

  ○ **Actionable Rule:** Engage in one new playful activity each week to discover new interests and skills.

It turned out that **the AI only considered the most recent list of rules it had created**. It seems **I waited too long to**

create the next prompt and found myself in a new session that had no way to refer back to my previous rules. This illustrates one of the problems with today's generation of AI. **Each session is independent of all other sessions.**

There is no "memory" kept between sessions.

Also, the input limits of 4,000 tokens (roughly equivalent to words) create a serious limitation in the usefulness of today's AI.

So, I'm now going to try copying from my notes into the AI chat window with all of those earlier rules. This could be a problem since the AI is limited in the number of inputs it can accept. So, let's try it and see what happens.

The first attempt will be to see if I can copy the entirety of the existing rules into the chat...

Success! Here's the first part of the prompt. It took almost all of the 4000 tokens, so this is not the best approach when you need to cite larger sources, but it worked for this instance.

"Please refer to the information below as source material to rewrite and score as you did in the examples above:

- Child
- Teen
- Young Adult
- Marriage
- Career
- Parenthood
- Middle Age
- Golden Years

Here is a list of 30 professions across various categories:

Project Manager"

The resulting list was quite long so I'll show just a couple of examples from it:

Golden Years:

- **Stay Active**: Engage in a physical or social activity several times a week.

- **Life Reflection**: Write down or share a life story or lesson learned with someone younger once a month.

- **Adapt and Learn**: Try a new activity or learn something new every few months to stay mentally active.

Now, let's move on to the professions:

Project Manager:

- **Goal Setting**: Start each project by clearly defining and communicating its goals.

- **Risk Assessment**: Conduct a risk assessment at the beginning of each project phase.

- **Stakeholder Updates**: Provide regular project updates to all stakeholders.

Next Steps

I have a couple more things to accomplish. One is to have the AI review this post and suggest edits. [I have incorporated those edits already] Finally, I'll ask it to produce some visuals to accompany the story.

My first two attempts resulted in this message:

Oops! Try another prompt

Looks like there are some words that may be automatically blocked at this time. Sometimes even safe content can be blocked by mistake. Check our content policy to see how you can improve your prompt.

But I finally got some nice illustrations and chose this one.

"Illustrations by Dall-E."

That's My Perspective...

01-08-24
AI Burnout

Is AI Just a Rabbit Hole?

All is not well in AI world… at least for today.

As I've been working my way through using AI to create an entire article, I find it impossible to stay completely on track. Each step of the way, I'm finding things that exhaust my capacity, divert my attention, and slow down my writing process. Here's a quick summary of the challenges from the past two posts.

Getting the **illustrations right was surprisingly much easier than I expected**. The AI created some examples based on my prompt, I selected from them, made a few minor changes, and got great results I couldn't have had any other way. Kudos for that!

"Illustrations by Dall-E."

Generating lists of potential topics was a breeze. After a couple of tries, I had my lists, which I copied into my note-taking tool. This sort of brainstorming assistance seems to be a sweet spot for ChatGPT. I thought the rest of the project would be a simple exercise in documenting the process. **My confidence outran my competence…**

I have a huge list of potential "Rules of Thumb" I want to evaluate, but the day was winding down, so I decided to return later to finish.

Starting in a fresh session the next day, I began to **immediately hit seemingly small obstacles.** First, **none of the preferences I had set up before the previous session had been retained.** Re-establishing my preferences each session is a small distraction that I'm certain won't be needed soon.

Then, Then, I found that ChatGPT couldn't read the format of my notes. So, I had to export them to four massive .pdf files and then convert them to text. However, the **files were far too big** to put into the chat window since it had a limitation of about 4000 tokens (roughly 4k words). So, I needed to **break down my notes into sections and manually cut and paste them into the chat window.** How's that for intelligence? This tedious and very repetitive process can't be automated in ChatGPT (yet).

I decided to **try a different AI called NotebookLM**, which could directly use my notes as a source. There's a learning curve to master the new AI, so that took a couple of hours to set up and restart with my notes now being direct sources for the NotebookLM.

This AI can't read any of the prompts I made for ChatGPT, but those weren't working all that well yet anyway. Let's **create a new set of prompts…** Wait, what? **Illegal word**

error?? Maybe I made my requests too complex? I'll start with a simple index of the words in my notes. I had to split my notes between 4 different, very long pages to create the .pdf files that were then converted to text files. I get an index, but it only includes words from the first source file. Simple fix... just don't forget to check the little box next to each source file name. Okay, I'm all set... I've checked the boxes and added just one tiny thing, "please also list the source page name with each item."

Drum roll, please! ... Wait? What!? You can't provide page names because the pages aren't numbered?? Okay, number the pages, resave them, and try again... Now it looks promising until I realize there are a LOT of words in this index that aren't part of the source files at all... I tried the obvious and asked the AI why those words were in the index, and it made up an implausible explanation that boiled down to "They seemed like good words to put into an index."

Okay, I think I've got this. These are **the "hallucinations" they mention in the warnings**. Rather than tell you about only your source materials, the AI includes things it has gathered from other sources.

By this point, I've spent about 4 times longer than normal on this article and stretched it out over three weeks. I'm almost giddy with all the distractions and having a bit of a hard time remembering what my original objective for this article was. I found **a few things the AI did amazingly well** and **a lot of things that require much more careful instructions to the AI** than I can deliver on the fly.

I gave up on using AI end-to-end for my workflow for now. I'll use it more as I get more familiar with what works well and what to do on my own.

Here's my bullet point run-down:

- Astonishing illustrations!!
- Topic and idea list wizard!
- Distractions abound.
  - Error messages
  - Size limits
  - File type constraints
  - Bogus instructions and completely hallucinated answers
- Errors compound and get replicated until the next session
- Partway through a session, the AI "forgets" what's already been done and has to be refreshed.
- Massively entertaining and **almost good enough** — until you look closely
- It can't explain its logic or reasoning coherently.
- MASSIVE TIME SUCK
- **Damn, it's a lot of fun though!**

Is AI worthwhile? For some things, the answer is HELL YES!

For others, at least for this week, (Things are changing at incredible speed.) the answer is HELL NO! but that will change.

In the meantime, next week's post will have (mostly) human-generated content, and I'll use AI to assist with illustrations and the things it does better than me.

That's My Perspective...

01-15-24
Taking a different look

Instead of forest, tree, weeds, and rabbit holes, let's zoom out a bit.

"Illustrations by Dall-E."

In my last post, I went way down the rabbit hole of exploring AI as a tool for writing. When I started this series, I hoped to communicate some important insights about making great decisions that are unexpected and worthwhile. Sharing my experience with AI as I created the article was meant to be informative and fun, but it derailed my thought process's original purpose So, let's zoom out a bit and restart with a different perspective.

Some True Things

The decisions you make determine the life you'll get. Better Decisions = Better Life.

No one has unlimited reserves of time, energy, or resources. If you want to make the best use of your efforts, focus on the things that are in your control or that you can influence.

Three major things influence results.

1. **Random Chance:** Random things happen that affect your results. To improve your results with randomness, make choices that **take probability and risk into account.**

2. **The Choices You Make:** Everything else being equal, better-quality choices have better results.

3. **Starting Position:** If you start with money, education, a great reputation, and the ability to learn, your results (the outcomes of your decisions) will be better. **Every choice you make affects your starting position** for the next decision. This is a cumulative advantage (or disadvantage if you make bad choices).

   Starting position matters and can be improved. Seek ways to improve your current situation. This could mean building savings, getting insurance, keeping up with your vaccinations, exercising, learning, flossing, or improving any dimension of your life.

   Some choices have an outsized effect on our lives. Finding and focusing on those choices where your efforts will be most effective makes the best use of your limited energy.

   Small changes in either a positive or negative direction accumulate over time and lead to better or worse outcomes. A small change at the beginning can result in a large difference at the end.

Every choice you make affects your starting position for the next decision. A better position might mean having more money, a better education, or just being the person more likely to be picked for a project.

Where to Put Your Efforts

The **big, impactful choices** like who your spouse will be and what career to choose **always** deserve a serious decision process (and the work that goes into it). **No extra effort** is needed here because you're already paying close attention to the critical stuff.

The **trivial decisions**, such as where to eat dinner, have no long-term impacts. You can **improve those choices a bit by making yourself a personal rule and pre-committing** yourself to a rule to avoid desserts except on weekends, for example. Commit when you're not tired or hungry. Still, don't waste effort on these. Flip a coin and get on with life.

"Illustrations by Dall-E."

Most of the day-to-day choices we need to make fall well below the bar of being life-altering. Some don't matter at all in the long term. Choices that are **hard to identify as important** can have big impacts over time and usually fall into the category I call "**Everything Else**".

How do we decide what to spend efforts on?

We can't know in advance, but we can minimize the effort we put into each choice, put up guardrails to prevent disaster, and improve the quality of each choice.

First and most important: Take a moment, step back, and **triage how much effort it's worth**. Here's a suggested way to sort:

1. **Trivial choices** — **No long-term effects or easily reversible**. Just flip a coin, go with your preference, or pre-commit to try something new every once in a while. **No effort** is needed.

2. Life-changing, important, consequential, permanent, or high-impact choices — This sort of problem requires a full-on conscious decision process.

3. **Easy choices** — Any choice that has a single, obvious best approach. These are the "no-brainer" choices. The only real work here is to **put up safeguards** (to notify you if conditions change or results aren't what was expected) and **reduce identifiable risks**.

4. The middle ground — Not high-impact or particularly risky but **not with a clear best choice**. This is the majority of decisions we need to make. It goes into the "Everything Else" sack.

5. Habits and small choices — What we eat, how we exercise, and many small, repeated behaviors. They go into the "Everything Else" sack too.

The first three types of choices are simple to sort. For our "Everything Else" category, we're left with those choices that have **some impact, may have cumulative long-term consequences,** or where **there's no way to know** what you need to make an "optimal" choice. These are worth spending a little effort on, but not worth the deep dive of a major decision process.

You've just done one of the two hardest things about any decision. You have s**tepped back from it for a minute to consider how impactful it might be**. You're off to a better start already.

Everything Else

Here's a thought experiment. Imagine you can make a small improvement in decision-making that results in a slightly better outcome on average. **What would that do to your life over time?**

If you used that improvement **on trivial decisions, it wouldn't mean anything at all.** Making better choices about things that don't matter over time is a waste of effort.

If you used that **improvement on the "MAJOR" decisions only?** You're already aware this is a major decision. You should be doing a full-court press on it.

A very small boost on the "Everything Else" decisions would result in improving most of the choices you make, resulting in more gain for less effort.

But no one in their right mind would want to obsess over things like habits or small day-to-day choices…

… unless there's an easy way to make better choices without wasting time or effort.

In our thought experiment, we've looked at which type of choices to focus on for the maximum impact. The "Everything Else" and "Habits and small choices" categories make up most of the load. Putting effort there is going to be worthwhile if we can find **reliable ways to make those choices better** without having to process everything individually.

This is where a "bulk processing" method to improve decisions is useful. This is where you'll find that great "Rules of Thumb" can be your superpower.

3 Rules to Start With:

1. You are responsible.

2. Look before you leap (Triage).

3. Don't do stupid! (Risk Management).

That's My Perspective…

01-22-24
Picking Rules To Live By...

The point of this post is to put great habits and rules into the mix.
To do that, here's how I'm choosing them.

In my last post, we looked at positioning and how selecting great rules of thumb can help us to always move toward better positions without wasting effort where it won't do as much good. I picked three examples to start with. Let's dig into these three a little further before we get into how to select your rules.

1. **You are responsible.** No matter what happens, whether it's due to your planning or to random chance, accept that you have full responsibility for your actions and your reactions to events. This puts you in the frame of mind to take advantage of opportunities and overcome obstacles.

2. **Look before you leap** (Triage). Step back. Take a breath. Now take a moment to triage this decision (or problem). Is it impactful to you or others? Will it have long-lasting effects? Deciding this tells you how much effort the decision deserves.

3. **Don't do stupid!** (Risk Management). If a choice has a 95% probability of success, it has a 5% probability of failure. Failure will happen one time out of 20. That's a great bet if you can afford to lose the money. If losing can bankrupt you it's a sucker bet. Always play to stay in the game. This includes bets on money, friendships, health, or anything else. The game isn't fun if you're not alive to play.

I picked these rules as examples only. There are thousands we could examine, and this is a good place to start.

What Criteria Should I Use?

I have collected hundreds of unexamined rules of thumb, but some are generic, not broadly useful, contradictory, or can't be acted on. In an earlier post, I mentioned some criteria for selecting only the most useful ones.

Here are my current criteria for deciding which rules I'll try to keep. No single rule covers everything, but any good rule should touch on most of them. I look for a rule to be generally applicable and it should be easy to remember.

1. Does this rule reliably improve my chances of a good decision? Is it better than a coin flip? An example is taking investment advice from a broker. They have a worse return on average than chimps throwing darts, and they make their money even when I'm losing mine. (Perverse incentives are a bitch.) I can reliably do better by picking a stock index fund, a bond index fund, and other investments. If the rule doesn't pass this first test, discard it or find a better one. No need to proceed unless it passes this test.

2. **Does it position me better** for the future? Good positions create options, while bad positions reduce them. You don't have to be an expert decision-maker to get better results, you need to put yourself in a better position. What a better position means can be hard to define but having skills, savings, education, and health are examples.

3. **Can this rule be acted upon or used as a trusted guide?** Otherwise, it's just an aspiration or perhaps a direction to travel. "Go big or go home" is an example of a useless rule because it has no sensitivity to context, just blowing sunshine. Similarly, "The nail that stands up gets hammered down." is just an encouragement to blend in. Neither is appropriate in some cases. A better example is: "**Observe, Orient, Decide, and Act** (Called the OODA loop).

4. Does this rule **encourage me to think through more alternatives?** Does it help find possible win-win solutions? Sometimes the best choice is to simply do nothing. Most choices aren't simple binary yes/no choices.

5. Does it **minimize harm or risks** to me or others? Running a pilot project is an example of risk management, setting up guard rails is another example.

6. **Is it prosocial?** Hurting others hurts me too.

7. Does this rule **fit into a systematic approach** to making good decisions in general, even for trivial things?

8. **Can I make this a habit** so that my default mode is to make the right choice?

9. **Does it increase my learning, skills, or understanding**? Does it help me learn from my own mistakes? Examples: reflecting on the day's work. One of my favorites for years has been "Follow the Money". It helps me understand what incentives people have and lets me respond knowing they have personal agendas.

10. **Is it a rule I'm willing to pre-commit to?** Some rules will have exceptions but the mechanism of precommitment means that any rule I've committed to will need a solid reason to override. It also means I will need to review that rule to decide if it should continue to stand. I can commit to a certain course and put that into practice until it becomes my nature. This final rule turns a guiding rule into an appropriate set of actions and, eventually into habits.

Which way?? On one hand… but on the other hand… and the other one…

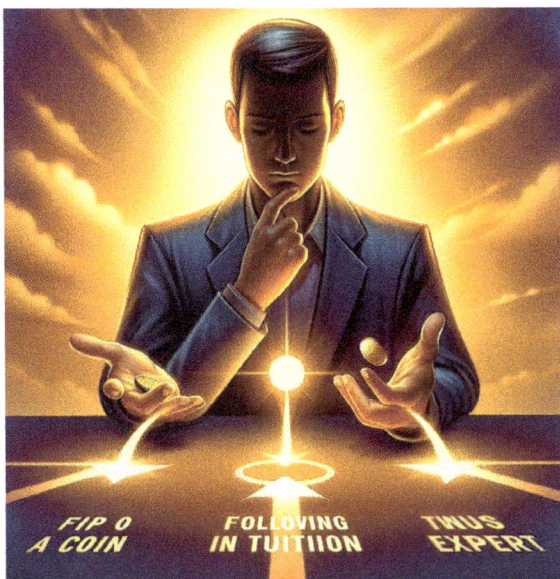

"Illustrations by Dall-E."

I should mention here that I'm just like most people, that is to say, I'm pretty lazy. *No way in hell* I'll work through all these steps whenever a decision needs to be made. So, I plan to filter my "candidate" rules (maybe with an AI) to eliminate the least useful ones and organize the rest into a cohesive "Operating System" for myself. Also, no way am I going to drag you along through that entire process, so I'll just present an example.

For example: "Thou shalt not kill", which should be more accurately stated as "Thou shalt not murder." seems like a good starting rule.

1. Will it reliably improve my chances of a good decision? Murder is rarely the best choice that can be made.

2. Does it position me better? Yep, better to not be on a wanted poster.

3. Can this rule be acted upon? No, but in this case NOT doing something is entirely the point.

4. Does this rule encourage me to think of better alternatives? Yes. It doesn't offer much advice about what to do instead but that leaves open a very wide range of actions.

6. Is it a rule that helps or minimizes harm? Check.

7. Can I make this a habit? Pretty sure that is possible.

8. Does it increase my learning, skills, or understanding? No, but this rule is batting way above average already.

9. Does it reduce risk? Obviously

10. Is it a rule I'm willing to pre-commit to? I have no problem with killing to defend someone, including myself, or in defense of my country, but murder is off the table. So yes. This rule easily makes the Operating System Handbook.

Strategy first. (Actually, done second in this case, but whose post is this anyway?)

My overall objective is to be the best person I can be and live my best life, according to my interpretation of what those are. Having an **objective without a plan to deal with obstacles is just an exercise in magical thinking.** So, I ask myself "What are the obstacles and what is my plan of action to address each one?"

- Human nature — my own biases, weaknesses, blind spots, positioning, and habits.

  o For important, permanent, or very impactful choices, get good at Decision Science. Learn to triage decisions so the important ones get proper attention. Think like a scientist.

  o For everyday choices, ***create a set of rules*** to act as guides and make it a habit to follow those rules unless there's a compelling reason. Get into the habit of doing the right things.

"Illustrations by Dall-E."

Random chance — How will I take advantage of luck, whether it's "good" or "bad"?

- **Position myself well** and always be looking for ways to improve my position on every important dimension such as health, relationships, finances, or learning.

- **Learn about probability and risk management**. Prepare for things to go sideways and be ready to help the people I care about.

Strategy

- The laws of the universe — **No magical thinking.** Reason from first principles but keep second-order impacts in mind. When you run into things you cannot affect, think of those as "gravity" problems. You can't change the law of gravity, but you can still find ways to fly.

In this post, I tried to share what I think is a good approach to selecting the "rules" I will plan to follow. I'll be making a pre-commitment to myself and working to make the rules I pick an ingrained part of my character. I chose a basic **strategy** (objective, plus direction to reach the objective, plus plans to deal with obstacles) and starting **criteria** for my rules. Then I ran an example rule through the filter of those criteria.

Note: My editor reviewed this post and commented "No one puts this much effort into making decisions. You don't even do this yourself." She was right. I only put in the heavy lifting and serious decision science for serious and impactful cases. For the everyday challenges I do a quick triage to make sure that I am treating the choice appropriately and then, assuming it doesn't have the potential to really foul up my life, I rely on my personal habits, experience and rules of thumb.

That's My Perspective...

01-29-24
Teaching,
The New Engine of Our Evolution...

Evolution never stops. It is accelerating!

In my last post, I shared something essential. It was the concept that, once you triage a decision and determine that it is not a career or life-threatening situation, **you can rely on great rules of thumb** to improve your decisions.

Don't Do Stupid. If it can hurt you or kill you or others, even if there's just a small chance, make a different choice.

Do Follow Your Own Rules. For the "Everything Else" choices, commit yourself to rules to help you have the best possible positions and outcomes. Make the in advance with great Rules of Thumb.

At the end of that post, I teased that this one would be about teaching.

When you have something exceptionally worthwhile to communicate, **teaching is the most valuable skill set you can have.**

We evolved over eons as changes slowly accumulated. Then, only an eyeblink ago, we added the ability to communicate. Evolution sped up from the pace of **one small change per generation to today's frenetic pace.**

Education is the second way we transmit knowledge, but it is much faster than evolution. For example, it might have taken humans hundreds of generations to evolve the ability to survive polio without any bad effects. It would have taken hundreds more generations to spread that immunity to the entire population. **With science, we developed and distributed immunity worldwide in a single generation.**

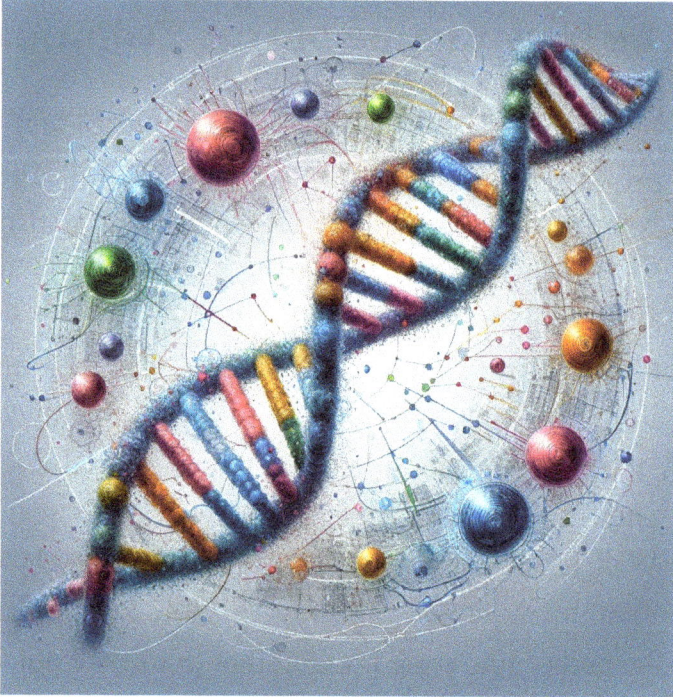

The new DNA

This step has allowed us to inhabit so many niches in the world is the knowledge we accumulate and pass on. Physical changes that take hundreds or thousands of generations can't compare with adaptations that take less than a generation. As communication technology (itself a product of science and knowledge) speeds up, **evolution has found a way to go into warp speed. Teaching is the key.**

The culture that embraces learning and adapting will be the culture that thrives.

Teaching is a process with the teacher acting as the enabler. Here's my perspective.

Major steps of the process:

1. Select and curate the information (curriculum).

2. Break it down into bite-size chunks.

3. Build scaffolding to help students grasp the concepts by tying them to existing knowledge.

4. Engage the students and get them thinking about the concepts.

5. Deliver or present the information in various ways, including having the students seek it out. (Get the information from the book to the students.)

6. *Students focus attention on the material and try to get a basic understanding. (Input)*

7. *Students practice with the material, connect it in different contexts, and gain proficiency in recalling and using it. (Processing)*

8. Find out what you can do with it. (Output)

9. Check for understanding and mastery. Evaluate progress to reinforce learning where it's needed.

> Notice that items 6, 7, and 8 are entirely controlled by the students, not by the teacher. **Learning is an active process.** The student must actively participate to learn.

> With all of that in mind, **can a teacher teach a student who is not engaged?** No.

> Can a teacher make learning easier for students? Yes, if they are trying to learn.

> Is it always a good thing to make learning easier? *****SURPRISE! *** The answer is No.** Here's why...

> "Desirable Difficulty" in learning means that the extra effort to process and internalize hard (but not impossible) things results in better understanding and retention. This

is because the extra effort reinforces the neural connection pathways that make memories, and we make more connections with other concepts, making this concept easier to recall in different circumstances. It seems that **getting the concept into the student's head is only part of the job.** She must be able to recall and use the concepts to master them.

Best teacher in the world

The best teacher in the world cannot teach an unwilling student, except with tremendous effort and maybe subterfuge. However, a willing student is an autodidact (self-teaching) in the sense that it **is up to the student, not the teacher, where to direct her attention and efforts**.

"It is far better to render beings in your care competent than to protect them" — **Jordan** Peterson.

So, it follows that if the student is engaged and trying to learn and accepts responsibility to "put the material into the brain", **the student can be a better teacher than even the best teacher in the world**. It also has some advantages because there just aren't enough top-notch teachers to go around, especially when it comes to things that aren't part of the curriculum.

When I teach, I make it my goal and explicit purpose to turn each of my students into people who can teach themselves. This means **teaching them how to learn for themselves, even when there's no teacher available.**

So, for example, when working with students to learn a new throw (I taught elementary students and adults the Japanese martial art of Aikido as well as teaching business and technical subjects like Project Management to adults), I'd ask a question to get them thinking about the throw and getting them engaged, then demonstrate it at low speed so they could get a fundamental idea how to move. That was the scaffolding or on-ramp to help them go in the right direction. Then, **practice, practice, practice.** Try it from different angles, with different opponents, at a very low speed.

"Illustrations by Dall-E."

This puts the information into the brain of the student, but mastery of the technique requires the student to truly understand the feel of it and **be able to fluidly and instantly call on it and use it in a variety of settings.** (Sort of like the mastery of basic math I wish all of them had).

As I got older, my body wouldn't allow me to demonstrate some things. From necessity, I've found that the best way to teach my students is to engage them and **let them teach themselves.**

I made it my goal to **teach them how to learn, not just Aikido, but anything they want to learn** by explicitly telling my kids that they were going to be the best teachers in the world (for themselves) and explaining my goal of teaching them enough that they could continue to learn on their own.

I know teaching kids how to learn for themselves was never explicitly part of the curriculum when I was in school. Critical thinking and decision science skills were not on the horizon. Discoveries of recent neuroscience (a word that wasn't coined when I was in school) are starting to yield better ways to grapple with teaching and learning.

There's lots of movement in the right direction. With people like the dedicated educators who are part of the "What School Could Be" community exploring better ways and working tirelessly to make truly effective education a reality.

So, to answer the question I posed in my last post **"Who is the best teacher in the world?" When teaching yourself,** you are. The only person who can put things into your mind, recall them, and use them is you. That leads me to the obvious next question, "Who is the best teacher in the world at teaching others?" I firmly believe that you can be that teacher if you can teach your students to learn "How to learn".

That's My Perspective...

02-05-24
Positioning your mind...

Working on getting my head in the right place.

The only true mastery is self-mastery. — Morehei Ueshiba (O Sensei)

In my last post, I called out teaching as a true superpower. **Before you can teach anyone, you need your mind in a good place.** Anyone who has ever taught children, teens, or adults will know that they test you (to your limits sometimes).

"Illustrations by Dall-E."

I've spent most of my 72 years trying to master my own emotions, especially my anger. I am not always the calm, caring role model I try to be. When a ten-year-old treats people with open disrespect in my Aikido classes, it's hard to stay calm. When an adult grad student presents a know-it-all attitude before ever working at a real job, I can get steamed up. When a colleague argues with me, I argue more passionately. I get defensive easily. I get **Angry** far too easily. This can be my default setting.

What do you mean wrong? Hulk never wrong!

Anger is not the place you want to start if you're trying to teach, coach, support, and build up others. It's also not a good place to spend time in general.

My temper got stupid too many times. When I was about 12 years old, I lost my temper because my younger sister could do a handstand pushup and I couldn't. I got embarrassed and flipped my dad's desk over in a fit of rage.

Dad had been a professional boxer when he was younger and knew something about dealing with rage. He taught me the error of my ways through a combination of hard labor (mine) and self-discipline (his). Among other things, he enrolled me in karate classes to teach me self-discipline which let me begin the work to master myself.

I have a lot of experience with anger, and here are some things I've found that help me.

- **Learn how to breathe.** Teach yourself to back off for a moment and take a deep belly breath in and out. Make it a rule and it becomes a habit. This slows down the chemical dump of adrenaline in your body and gives you a chance to react with some thought instead of blindly lashing out. Martial arts and yoga can help you, or Google "4 by 4 breathing" for a quick intro.

- **Exercise every day.** Physical health is a requirement for mental health! Sitting at a desk or watching TV won't let your body flush out the stress chemicals that can build up fast. Doesn't have to be much — a long walk or do some squats and stretching.

- **Get enough sleep.** Lack of sleep makes you irritable. Also, it makes you stupid. Get a bit more sleep than you think you need, your body and mind will both thank you. Naps are wonderful but not too close to bedtime. Avoid caffeine late in the day.

- **Anger is often triggered by fear.** Could be fear of losing status, having your ideas put down, fear of loss, shaming, physical fear, or anything that your mind interprets as a threat. Learn if the things you fear are real or exaggerated.

- **Talk with someone you trust.** It's a sign of strength to have the confidence to face your problems and deal with them rather than denying anything's wrong.

- **Be social.** Maintain your friendships and spend time with people you like.

- **Get help.** Get help. Don't be the fool who won't see a doctor until the cancer has already killed you. You can do positive things on your own for mental health but sometimes a band-aid you can put on just isn't enough. Don't be embarrassed. Would it bother you to see a doctor for a broken leg? Broken emotions are just as serious. **Get help.**

- **Stop doom-scrolling and cut back on the news.** Media has a huge incentive to capture and keep your attention. They focus tightly on the horrific, terrible, scary, and unusual. Focus on the things affecting you.

- **Drugs can be a help.** Your brain is living in a bath of chemicals. If they get out of whack, so do you. Don't hesitate to use them if your doctor prescribes them.

- **Learn to tell the difference between things you can impact and things you have no control over.** If you absolutely MUST be angry at something, reserve your anger for the things you can change — then use the anger for fuel to do it.

- **Meditation can help you separate your feelings from who you are.** It can help you step back a little and realize that this emotion is a passing thing that you can choose to ignore or choose to lean into.

- **Work with your hands.** Play a guitar or repair something around the house. I don't know why it works, but it seems to help keep me calm and the effects last for a while.

"Illustrations by Dall-E."

No guarantees but it seems that using these helps me. For me, anxiety leads to irritability, which leads to a short fuse, which leads to blow-ups over trivial things. Stopping that spiral as early as possible is my goal. The farther into the spiral I get, the harder it is to calm down.

Whenever something hits me (literally or emotionally) I make it my first rule now to **step back and take a deep breath... or several, before I react.**

That's My Perspective...

02-12-24
Checklists, frameworks, and templates
Oh My!

Don't Do Stupid: Checklists, Frameworks, Templates to Avoid Ruin

"The checklist is one of the most powerful tools ever invented." That's what Atul Gawande, a surgeon and a writer, said in his book *The Checklist Manifesto*. And he is right. Checklists, frameworks, and templates help avoid stupid mistakes and omissions. In this post, I'll share one of my favorites."

"Illustrations by Dall-E."

"It is remarkable how much long-term advantage people like us have gotten by trying to be consistently not stupid, instead of trying to be very intelligent."

— Charlie Munger

DON'T DO STUPID!

My last post discussed getting my head in the right place and some effective tactics. Today, I'll share one of my favorite tools. It's a checklist I used to help with major decisions with my former project teams. Let me know if you find this useful.

Decision Checklist:

This decision checklist provided me with reminders for complex projects with a large global team. It's overkill for small decisions so modify it as needed. **Good checklists are precise. They are efficient, to the point, and easy to use even in the most difficult situations**.

Before you start:

- Eliminate the things that always make people stupid.
- Hunger, Anger, Stress, Fatigue, Haste
- Know your purpose, your "Why".
- Limit your losses:
  - Triage to determine the importance, urgency, and impact of the decision.

Define the Problem Clearly:

- What is the problem to be solved?
- Why is the decision needed?
- Define what an acceptable outcome will look like.
- Write out a clear problem statement.
- What information is required for a good decision?
- What does failure look like? Define it.

Create Options: Having more than just the yes/no option leads to better choices.

- Consider the "Do Nothing" or "Wait" options.

- Invert the problem: Instead of "Increase sales" try "Fix what is hurting sales."

- Get diverse perspectives. — Include experts, users, and customers.

- What would this look like if it were going to be easy?

- Get ideas from outside companies and other departments.

"Illustrations by Dall-E."

Refine the Options:

- Consider 2nd and 3rd order impacts. What are the effects downstream?[1]

- Ethical and moral dimensions: Does it pass the sniff test? Is anything rotten here?

- Evaluate and deal with risks. Avoid the risks or minimize their effects.

- Plan for sustainability and resilience.
- What must be true for this option to have a successful outcome?
- Review with competent others.

Decide and act:

- Negotiate and bargain for buy-in from the team.
- Record the decision results and major factors influencing the decision.
- Review the decision process after each pass. What worked, what didn't, and why? Improve the process.

Sometimes simple tools are most effective. Checklists are one of my favorites.

Templates reduce the amount of work and organize it better. Put your heavy lifting into the planning for your mental tools so you can repeat your successes and make sure you don't omit anything important.

Mental **frameworks** act as an on-ramp to get to the essential things right away.

Next time, we'll look at an overall framework that helps me guide my thinking.

A first-order impact example: Company X gives a pay raise to all its workers. A second-order impact would be that the workers would spend more in the local economy. A third-order impact might be some new retail stores open in the area. Sometimes the downstream impacts of a decision can be larger than the first-order impacts.

That's My Perspective...

02-19-24
The Big Picture...

Zoom out to get an overall view and find signposts!

In my last post, I covered one of my favorite tools, the humble checklist. It turns out to be extremely effective at helping to prevent stupidity and errors of omission.

This week's post explores a framework to connect ideas and approaches while keeping me from straying too far off track.

"Illustrations by Dall-E."

Start with your WHY: Know your values and purpose so you have a compass to guide you. You don't have to know the destination as long as your compass points you in the right direction. If you're trying to solve a problem, know why you need to solve it.

Keep your head on straight: Do the maintenance your body, mind, and spirit needs. Avoid making stupid mistakes because you're tired, rushed, stressed, or sleepy.

Always be moving to a better position: Three things contribute to the outcome of any decision you might ever face. They are:

1. Luck: Random chance and its effects. — Deal with luck by being alert. Prepare for both opportunities and risks.

2. Skill — The abilities you've built and preparations you've made over time. Have you done the work? Do you have the skills? — Always be learning and building your skills.

3. Starting Position — Compare the lives of someone with motivation and direction, money, health, education, and the wisdom to use them versus a person who is unmotivated, poor, sickly, uneducated, and directionless. — Always be moving toward better starting positions.

The Decisions you make interact with your Luck, Skills, and Starting Positions to produce the outcomes in your life. Each decision, from the smallest to the largest, contributes to the range of possible outcomes you'll have.

"Illustrations by Dall-E."

Better Decisions Lead to Better Lives.

Triage your decisions — Taking this step lets you focus effort where it's needed.

1. Big, important, life-changing decisions require your full attention. Use an evidence-based process to make the best possible choices.

2. Most decisions are casual and don't require a full-on decision process. That's the sweet spot for using well-curated Rules of Thumb. Put serious thought into the rules you want to live by before the moment comes when you'll have to decide. [More on that in my next post] …

3. Make decisions when you're at your best and create rules for yourself that you pre-commit to following. For example: Don't take anything you haven't earned. Another example: Only eat desserts on the weekend.

4. Pay attention to your habits. They produce a cumulative effect. Try to turn good choices into your default setting.

Great decisions mean nothing until they produce Actions.

Create Strategy — Make certain you clearly define the goal. Once you know the goal, anticipate obstacles and plan actions to overcome the obstacles. Without a goal, an understanding of the likely obstacles, and a plan to overcome them, you don't have a strategy, just an aspiration or maybe a general direction.

Act — Break your goal down into separate actions or tasks. **Assign ownership** of the satisfactory completion of each task each task to a single person, even when that person is just you. **Monitor** each task until it's complete and accepted. [I'll have much more on specific ways to make sure you can complete the tasks in a later post.]

Learn — The scientific method is the best self-correcting mechanism we've ever discovered. The key to self-correction is to make empirical evidence the deciding factor when judging any theory. In other words, does it pass the test in the real world, with skeptical reviewers closely watching in controlled conditions?

Review your process: Capture your decisions, the alternatives, and the reasons you chose each action in . When you know the outcome of a decision, review it. What worked? Double down on that. What didn't work? Why? Address that and adjust your process with the information.

Of course, a blog post can't go into much depth. I haven't touched on design, systems thinking, mental models, risk management, forecasting, tools, or other important considerations. We'll save those discussions for later posts.

That's My Perspective...

02-26-24
Downsizing again

Taking downsizing to a comfortable conclusion...

Downsizing can be a shock to the system. A few years ago, as I was retiring, my wife and I realized we'd need to downsize. We had raised a mixed family with four boys and were empty nesters. Our five-bedroom house was just too much. We downsized to a two-bedroom apartment while we intended to look for a smaller house.

"Illustrations by Dall-E."

We started the process early of selling things, holding garage sales, making donations, recycling, and sending some of it to the trash. We chipped away at it for almost a year before we sold the house.

We rented a nice apartment and moved in. The two-car garage was full to the ceiling with no room for cars, and a large storage unit jammed full.

If my life has taught me anything, it's taught me that plans change! When the housing bubble burst and interest rates rose, buying another house was impossible. This moved us into phase two of downsizing, which puts real physical limits on the things you keep. We needed to get rid of more stuff.

Phase two: In the first phase, it was simple, get rid of stuff you don't like. In phase two, we had to set some stricter rules for ourselves.

Downsizing rules:

- **When something comes in something has to go out.** We carried it a little farther for a while and got rid of two things for every new one.

- Set aside a staging area for To Sell, To Donate, or Trash. Do NOT get rid of something your partner wants to keep. It helps to have a staging space so you can both look it over before it goes out.

- **Sell it if you can**, donate it, recycle it, and junk it only if you must.

- **Feelings matter.** Some things feel as if you're throwing away a part of yourself. I have a soft spot for the tools my dad left me, even though I don't use them anymore. I'm keeping a few.

  - You don't have to get rid of everything. If an object brings back good memories, keep it. **Downsizing doesn't mean you stop living.**

o As you tackle an area, **stick to a small space at a time**. It's draining to have to make a lot of decisions all at once. Instead of "clearing out the garage", try sorting through the tools in one toolbox.

o If you're keeping something because you *might* need it, ask yourself if this thing can be rented or borrowed.

> ➤ I had a lot of tools used only occasionally. It boosted my self-esteem to think that I could fix almost anything but…
>
> ➤ I'm not doing any more remodeling, not because I can't but because I live in an apartment. Now I realize that I'll never lose the knowledge and I can get the tools if I need them.

- Except for foul-weather gear and emergency supplies, if you haven't used it in a year, it goes.

- Be pretty ruthless. If you're like me, you won't miss something once it's gone. We had stuff in the rental storage unit for almost a year and realized we couldn't even remember what was in there. That made it easy to get rid of the unit and the monthly rent.

- Stay at it in a session until you can see some visual evidence of less clutter. The good feeling as we reclaimed the space for the car in the garage boosted our efforts and morale. Small successes feed on themselves.

We had a deadline to meet with the move to our apartment. That was the easy phase.

Phase two was more difficult with tradeoffs such as missing the occasional fresh bread from the bread machine but appreciating the space it opened even more. **Once you realize that the things you have don't define you as a person, it starts to become a challenge to see what else can go next.**

Now, we're no longer specifically downsizing. We **continue to declutter** and get rid of excess things because it makes a nicer home. No pressure to get it done before moving day makes phase three much more comfortable.

I'm making another pass at my closet next. It's a process.

"Illustrations by Dall-E."

That's My Perspective...

03-04-24
Following the logic...

Taking a simple rule to see where it leads us.

Let's explore some results you'll see if you look closely at the rule that drives evolution, "Survival of the Fittest".

It's the mechanism behind every living being. "Every living being must adapt and reproduce or die." It's elementary.

"Illustrations by Dall-E."

It doesn't set the bar impossibly high. If you can find a way to get to sexual maturity and successfully reproduce, your children will have the chance to survive. That leaves a lot of room for experimentation. All living things are running that experiment every day.

Imagine you were one of the very first living beings on earth, billions of years ago, a collection of cells with one purpose — reproduction. That means you had to survive long enough to reproduce. Maybe you have a mutation that allows you to move slightly better than others toward food or away from danger. In a few thousand generations, your offspring will fill the environment. Back then, the bar for survival was low. There was little competition for scarce resources.

Think of surviving long enough to reproduce as the ratchet mechanism on a jack. Once the ratchet clicks, the advantage you have passes to the next generation.

RULE #1 — Survive

Competition enters the picture once a resource becomes scarce and more than one individual can access it. Other living things are competing for resources. Competition is one of the pressures driving evolution. Think of it as one of the forces pushing down on the jack handle.

Rule #2 — Survive AND Compete

We are all competing for resources. It might be food, available mates, cooperation with the group, status, or any other resource. At the same time, we must avoid things that injure us. More pressure to apply to the jack. The process of evolution speeds up.

Mother Nature is running a risk-and-reward experiment. We are it. She keeps leaving her bets on the table and our evolution is the payoff. Over time, following simple rules allowed evolution to produce us all.

Rule #3 — Culture and education change the rules.

Think of them as the hydraulic lift, or maybe more appropriately as a rocket booster!

Imagine you've come up with a new survival trick that helps people thrive. That idea can become a meme that spreads through the culture (versus thousands of generations to spread the evolution of a genetic feature.) Of course, your meme will have to compete with all of the other ideas in circulation and survive to get broad exposure.

This is an example of evolution in action. Learning and spreading the best "memes" allows those who can adopt them to thrive and compete better.

"Illustrations by Dall-E."

Here's a practical example you can try on for yourself.

Dealing with "stuff" — Filter and Focus:

Pay attention to the things you can change. Only a tiny part of all the "stuff" that comes at you is something you can have any effect on. You probably don't have the answer to war, famine, disease, or tragedy no matter how much you care about them. Any effort you spend on the things you can't affect is a waste of time and energy. So, it makes sense to pay attention to the things you can change.

"Illustrations by Dall-E."

Ignore the stuff you can't change!

There are too many things that you COULD change to be able to do them all. So, it makes sense to **focus your efforts where they will help most.**

Pay serious attention to the important stuff. Some "stuff" is life-changing, irreversible, or high-impact. However, most of us only face these kinds of decisions a few times in our lives. So, it makes sense to give them your full attention when they come up.

Most of the "stuff" (almost all of it really) that comes at us is routine, every day, not earth-shattering. That's where we spend the most time. It is the small choices like "What to eat?", "Who to see this weekend?", and "What to do?" that make up most of our lives. We make most of those decisions habitually and easily. So, it makes sense to spend some focus and effort on making sure the habits are good ones.

"Illustrations by Dall-E."

You can't know in advance what stuff is coming — but you can make your system of small habits just a bit more likely to help you thrive. Small changes in areas where you constantly make choices will have a big impact over time.

So, if you're not currently facing an earth-shattering crisis, it makes sense to focus on a single, small habit and make sure that your default setting becomes one that's going to help you thrive.

Lather, Rinse, Repeat.

That's My Perspective...

03/11/24
One useful thing to unpack...

Be lazy — big results from little changes.

What's the big idea?

More than ninety-nine percent of the choices you'll make in your life are ordinary everyday decisions.[1] What mattress should we buy? Where should we go for vacation? What will I have for dinner? These are choices that aren't permanent, irreversible, or with big, immediate impacts. They are important, just not right away.

Small, easy changes to your "default settings" affect your life more, with much less effort, than improving on "Important" decisions. — Why?

1. The "Important" decisions already get your full attention.

2. Following a good decision-making process gives you outcomes that are hard to beat.

3. Small changes to the "Everyday" ninety-nine percent of choices will have more effect over time than changes to the "Important" one percent.

    Your "Everyday" choices are driven by habits and emotions, not logical reasoning.

    Think of your emotions and habits as "Default Settings". Imagine a picture of them as an elephant. [2] This is your subconscious mind.

    - Then picture the logical part (conscious reasoning) of your mind as a small monkey riding the elephant. As long as the elephant is happy, the monkey can gently suggest where it should go.

- However, if the elephant wants something else…

Small changes to make my "Default Settings" better can have a big cumulative impact.

✴✴✴✴✴

Why does it matter? — Positioning

Good habits and "Rules of Thumb" improve your positioning.

"Illustrations by Dall-E."

"Positioning" means your starting point. What position are you in at the start of each decision? **A good starting position helps get good results.** For example: Having great health, money, education and networks of friends makes it much easier to succeed than starting with poverty, malnutrition, or other disadvantages.

"Resources" means the resources you can bring to bear, whether that's skills, education, energy, friends and their contributions, money, time, or other resources.

"Luck" means random chance. It is the combined interactions of everything affecting your results.

Positioning + Resources + Luck = Results — Always be improving your position

✳✳✳✳✳

You have no control over random chance, but having more resources and better positioning helps improve your results. It follows that you should always be improving your position.

"Luck favors the prepared mind." — Louis Pasteur

I've talked about "Rules of Thumb" in earlier posts. You pick the rule to follow and make a commitment to yourself (and others) to follow it. If you use good criteria to pick the rules, you change your life's direction.

"Illustrations by Dall-E."

Rules of thumb are useless unless you can consistently follow them. Once you commit to a rule, you must make it a habit.

How to create a habit loop

1. Identify the mechanism that sets off an existing habit: e.g. smelling coffee makes a smoker want to light up, or a food commercial makes you want a snack. These are "triggers".

2. Identify the "behavior": e.g. smoking, snacking.

3. What's the "reward"? e.g. reduction in cravings, feeling of fullness.

That's it, a **trigger,** a **behavior**, and a **reward**. For example, when I brush my teeth, I automatically take my pills. The reward is just a check on my internal to-do list. Breaking old habits can be as simple as noting the trigger and substituting a different behavior and reward. — NOTE: I said "simple" not necessarily "easy".[3]

Use habit loops to steer your elephant

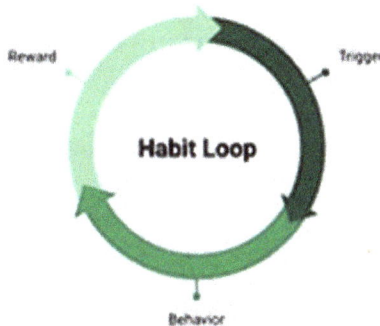

"Illustrations by Dall-E."

Remember that your subconscious mind already has hundreds of habit loops. You don't relearn how to tie your shoes each morning. These small loops can be directed where your conscious mind wants to go, but it takes time and repetition.

You can start new habits (or change existing habits) by noticing or selecting the trigger, creating a small, simple step to start the desired behavior, and rewarding yourself every time you do. Make certain that the small, simple step is **small and easy.**

So, if your goal is to run a marathon, make the **small simple step** of just putting on your gear and jogging a block. Your reward is a big smile and a high-five to yourself. If your goal is to lose 50 pounds, make it a habit to drink a glass of water before you snack on anything (and don't make more food the reward).

By changing your own "default" behaviors over time, you can gradually guide the elephant in better directions. Result — Happy Monkey!

1. Less than one percent of the choices you make in your entire lifetime will be critically important. When you choose a spouse, a career, where you'll live, you already know that you must put in the effort to get it right. Conscious, clear, and careful thinking is essential for your most important decisions.

A full-court press of careful, calculated decision-making is a lot of work and it's slow.

2. Thanks to Tim Urban and his "Wait, But Why" blog for the elephant and monkey metaphor.

3. James Clear's book "Atomic Habits" is an excellent resource.

That's My Perspective...

03/18/24
Not My Bucket List...

Working on My Kick-it List.

"Illustrations by Dall-E."

Seinfeld did a special when he turned 65…
It hit home.

Saying things like "**I don't have a bucket list.**", "No, I don't want to train for a marathon", "Why would I want to learn hot yoga?" "Not climbing any mountains, no skydiving either."

"No, I've improved myself enough thank you!" — *Jerry Seinfeld*

I don't have a bucket list, because I honestly feel that I'm doing more interesting things with my life than skydiving, learning a foreign language, traveling, visiting another sandy beach, climbing a mountain, sailing, or learning hot yoga. Nothing against any of those things but **they aren't my life's ambition**.

I just want to be a good person, help where I can, and leave things better than I found them.

So, I've set out to create a sort of Reverse Bucket List and **start getting rid of things that are holding me back.** Get rid of the hurtful and bad first.

Kick It List

1. Drinking too many calories (mostly sugar)

2, Too much crap that I don't need. Get rid of everything in storage. When something comes in, something goes out.

3. Smoking. What exactly was the upside??? I never found one.

4. Not moving enough every day (Still a work in progress)

5. Wasting effort on things I can't control (I'm better at this but it's still a problem.)

6. Not getting enough sleep.

7. Making snap judgments about important things. (Figuring out what's important needs to be a part of this.)

8. Snacking too much.

9. Anxiety and anger, getting better but this is a long-term project.

10. Beating myself up over anything and everything.

11. Reactive habits that put me in a bad frame of mind to deal with challenges.

Before you try to tune up the engine, get rid of the dead weight. I'm going to do that with my own operating rules. The Kick It List is a part of my OS.

That's My Perspective...

03/25/24
Getting it done...

Warning: long post...

If you are a "Doer" this post is for you. You've made the best possible choices. Here's how to make it happen.

Why it matters:

All the planning in the world is useless if you don't act. You spend a lot of energy and time making the right decisions. Once the planning is done, you need to act effectively to carry out the plan. This is true for small projects like moving a department or planning a party as much as it's true for coordinated global infrastructure projects.

Why I can help:

I finished my career as a Senior IT Project Manager after having directed massive global projects for AT&T, IBM, Siemens, Wells Fargo, and other Fortune 50 companies.

My cousin, a hospital director, asked me how to plan for a department move. He's a phenomenal anesthesiologist but has no project management experience. The result is a crash course in the essential elements of project management.

Project Management for non-Project Managers (Working with a team of people)

When you are managing or creating a change, think of it as a project. Here are the key things you'll need to know to tackle a project such as a large department move or your kid's birthday party.

Planning the birthday party.

Tools — There are dozens of Project Management tools available online. However, the learning curve may not be worth the return. With that in mind, you'll at least need a copy of Microsoft Excel and a copy of Microsoft Word or the equivalent applications. If you are working in a team environment, you'll also benefit a lot from team collaboration tools such as instant messaging and collaborative workspace tools such as Webex or Zoom which allow you to conference, share work screens live, and use whiteboards together from your desktop. Trello is also a good tool for collaborating on projects if you have the budget for it.

"Illustrations by Dall-E."

Key things to know about project planning:

Planning is the step that will benefit you most. Don't skimp on the time and effort spent here. It will save you more time and effort than it costs. It's a great rule of thumb to "Plan slow and act fast".

Things go wrong. Understand this and know how to manage the risks to minimize the fallout. You'll want to set aside a contingency fund of at least 10% added to the budget after all your estimates are in. Why? Because estimates are best-case guesses.

Communicate, communicate, communicate — If the entire team knows what is happening, what should be happening, and why, they can help you to succeed. Make sure everyone has access to the information they need to understand what's going on in the project.

Make sure the team knows it's **your responsibility** to help figure out how to deal with friction and obstacles and **their responsibility** to let you know of upcoming issues as soon as possible. No surprises.

Break down the tasks into manageable chunks that can be assigned to a particular person or group which can be held accountable for the results. Get the people who will be doing the work to help you define the tasks.

1. The chunks (tasks) should be detailed enough to provide you with actual results that can be tracked at least weekly. Sometimes a task will take a day to complete and sometimes it may take multiple weeks. You need to be able to track tasks or an estimate of how much of the task is completed at least weekly.

2. Track each task to completion. A task is only complete when the entire task has been done. The person receiving the work

verifies that the work is done properly before the task is accepted.

3. Every person or group that gets a task assigned must understand the task thoroughly and they must acknowledge responsibility to complete the task as they have committed.

4. Every person in a chain of tasks is responsible for making sure that the work from the previous person is acceptable and if not, to notify you and the person responsible so the work can be fixed. For example, if a map of the new location is needed, the person getting the map is responsible for verifying it has all the required information.

5. When complete, each task must be reported to the PM. Problems should be reported immediately. Again, No Surprises. (Never shoot the messenger).

6. In your status calls, you need answers to these questions:

7. What is complete?

8. What problems, obstacles, or friction is coming?

9. How can I help?

10. The project manager's role is not to perform the tasks but to assign the tasks, make sure they are understood, communicate and manage changes, monitor progress and quality, and direct efforts to deal with problems.

11. Put the tasks into a rough chronologically ordered list (Excel, Word, Trello, MS Project) with at least the following columns:

12. Task name

13. Task owner (person responsible for making sure it's done and reported. Not necessarily the person doing the work).

14. Start date.

15. Due Date.

The person responsible for doing the work.

16. The person accepting work.

17. Dependencies (other tasks that affect this task — e.g. construction start has a dependency on permits).

18. Percent complete. (use zero for not started, 25% for started, 50% for in progress, 100% for complete OR just use zero and 100%. Either the task is done or it's not)

19. Task notes

20. In meetings cover at least these topics:

21. Status updates (including budget if you are responsible for tracking it). Each task owner should provide you with an update weekly.

22. Change requests — examine any proposed changes to understand the impact on the scope of the project, schedule, budget, and quality. Make sure that everyone understands that each change will be reviewed and either approved or disapproved due to impacts on the schedule, budget, or resources.

23. Issues and Risks (Issues = already caused a problem, Risks = potential to cause problems.) Track each Risk and issue in a log file that assigns an owner to track each, the action plans in case a risk becomes a reality, and the plan to get back on pace if an issue arises.

24. Action items — every issue should have action items needed to resolve it, an owner, and a due date.

Managing risks

Risk management is at the heart of project management. Things go wrong for every project and risk management is the tool to correct that.

Doing a credible job in **risk management requires the active participation of the project team**. It's one of the few parts of project management where you get multiple benefits for a single activity. **Walk through the tasks with the team and ask "What are the kinds of things that could go wrong with this task?**

Consider:

- As you get estimates for each task, make sure the owners spell out any foreseeable risks in advance. This makes them think through what could go wrong and gives you a more accurate estimate.

- Ask this key question: For similar projects or tasks, how much did it cost and how long did it take? (People unintentionally estimate best-case scenarios when they take on a task. Asking how long it took others gives a better estimate.)

- Get the project tasks clearly understood, communicated, and critiqued by all the key project team members.

With the team, actively look for possible problems **and also possible opportunities that will help the project**.

Make sure the project team not only understands what needs to be done, it also understands **how problems in one area affect the other areas.**

- Take the opportunity early in the project, when changes cost the least, to identify and fix potential issues.

- When "Things go wrong", you will have the plan to deal with them in the most effective manner possible, as early as possible.

- When things go right, you get the choice of going to your stakeholders and returning reserves from the project budget or bringing the project further under budget.

Ways to deal with risk:

1. Avoid it — Identify the risk early enough to avoid it altogether or avoid it as much as possible.

2. Mitigate — Find ways to reduce the risk such as back-up plans, having extra cash set aside in a risk reserve, and allowing extra time for tasks that will benefit from the extra time.

3. Share the risk — Purchase insurance to cover as much of the risk as necessary or get a sponsor to absorb some of the risk.

4. Accept the risk — Sometimes there will be risks that are low impact or low probability that you should simply allow for. Put aside a risk reserve fund (I recommend 10% if you can negotiate that) as part of every project and add a small cushion of time to the project. That way, when the small, low-impact risks become issues, you'll be able to absorb them and keep going.

5. Set aside reserves of time, money, and resources to help mitigate risks.

6. Go through each task with the team and capture these items:

7. What could go wrong? (the risks)

8. How would that impact us? Is it a low, medium, or high-impact item?

9. How likely is that to happen? Is it low, medium, or high probability?

10. Determine how much risk you'll tolerate (e.g. for low impact/low probability, just monitor to make sure it doesn't get worse. For medium impact/ medium probability, write a risk management contingency plan, for high impact or probability, assign risk reserves in advance and have contingency plans.)

11. Track risks in a risk register until they are no longer risks or they have been dealt with.

For risks that are large enough to exceed your tolerance threshold, set up a risk trigger and assign an owner. Have the team performing the task tell you what the action plan would be if this risk turned into an issue.

12. Assign an owner to track each risk until it is resolved.

13. Review the risks each week with the team and keep an eye out for new ones.

Executing the project:

During the execution, hold weekly meetings (at least weekly, more frequent if needed) to update status, track progress, surface issues, and new risks, deal with changes and action items, and **communicate**.

The purpose of the meeting is to **make sure the team is on the same page and dealing with risks and issues as they arise.** It is to coordinate efforts and can be used to dig into problems to resolve them.

Catch people doing the right things and point those out during the weekly meetings. **Track progress and share it with the team** so they know exactly where the project stands. Keep status posted in a place that's accessible for the entire team, whether that's a shared server or a community bulletin board.

Be prepared to **hold people accountable** in the meetings for their deliverables. Do this in a completely factual and fair way. Don't assign blame, find and fix problems. Try to find them before they become big problems.

Communicate early and often with the stakeholders as well as the project team.

Celebrate small successes and be certain to celebrate bigger ones.

When things are successful, find them, focus on them, and double down on efforts in that area.

Celebrate the small and large successes. Learn from your successes even more than from your failures.

Good luck!

That's My Perspective...

04/01/24
One Brick...

Get going and keep yourself moving with this one-brick trick!

Definition: One Brick...

Instead of a goal, create a process. Define a minimal chunk of acceptable quality work. Call this "one brick". It's a minimal bit of work toward a goal.

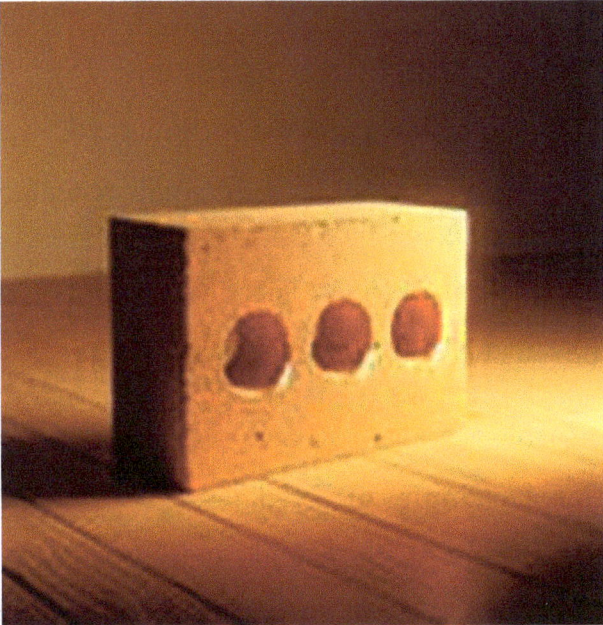

"Illustrations by Dall-E."

Defining a specific goal is needed for some projects. You have to be able to communicate the goal to others. You have to get everyone pulling in the same direction. The clearer you can get, your chances of reaching the goal improve.

However, **many important goals aren't definable at the start.** Getting an education, being healthy, building a community, connecting with friends, and having a great career are examples. It's impossible to clearly define the end product at the beginning.

Always move toward your goals. When a goal is something nebulous like "physical fitness", "writing a book", or any long-term goal that's subject to change over time, **a good process will help** reach it. Fitness needs will change, reaching a healthy weight becomes maintaining it, and the book you start evolves and improves as you create and edit it.

Building a process:

- Determine the general direction(s) you want to go.
- Decide on the "bricks", (the small, repeatable steps that mean progress).
- Repeat and Adjust as needed.

You may not need a specific goal, just a vague description like "fitness" will work fine, IF you have a clear direction and steps to take toward it.

Here are some examples of "bricks":

- **Writing** — Write 2 single-spaced pages.
- **Research** — Carefully read and annotate 3 chapters.
- **Exercise** — Do ten squats, pushups, and sit-ups, or do twenty minutes of walking and stretching.
- **Maintenance** — Empty the cat box and trash, make the bed, and make breakfast.
- **Education** — Read for an hour.

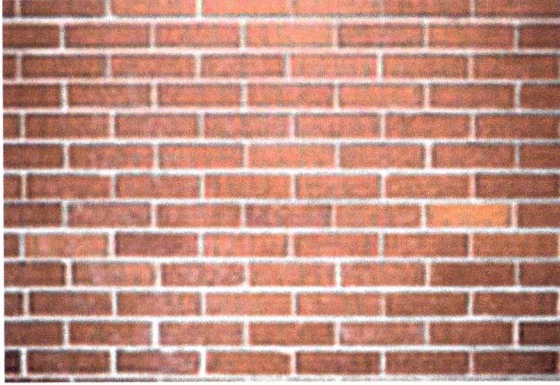

"Illustrations by Dall-E."

The one-brick trick will help get you past procrastination while moving you much more quickly than you realize. The effort is small and therefore not discouraging. You're always free to add more than one brick at any time.

Instead of a daily hour-long workout, pick a minimal brick and do that consistently. If you feel like more, do more. You're already in the gym clothes and have gotten a quick warmup at the very least. You're not going backward.

Two single-spaced pages daily will result in a 350-page book in about six months. That's a terrific pace for the "Great American Novel" draft.

- Writing this post: one brick.
- Illustrations and publishing: one brick.

That's My Perspective...

4-15-24
A Compass to Steer By

Navigating the day-to-day...

Problem:

It's impossible to keep all of the great advice and the accumulated wisdom handed down from centuries of successful people in mind. Yes, Warren Buffet has great insights into business, and ancient philosophers like Seneca knew a lot about dealing with life's problems. But it takes a lot of effort to pull up and sort through when it's most needed… Which is almost always when I'm hurried, tired, hungry, stressed, or at a low point.

"Illustrations by Dall-E."

Solution:

Use carefully vetted rules of thumb to manage most choices.

Here are some to consider:

- Triage your decisions. Give a moment of thought to how much impact this decision can have on you and others both now and cumulatively in the future. How will you feel about it in 5 days, 5 months, or 5 years? Can it be reversed easily or not? For high-impact, long-lasting decisions step back, take a deep breath, and give the decision your undivided attention.

- For all the rest, let your default rules be your guide. The huge majority of choices you make have a low individual impact but are IMPORTANT... That's because each choice changes your starting position for the next choice (and that compounds things over time). Choose the rules you want to follow in advance. Keep them simple and general. Making a tiny improvement to the rules you follow for everyday choices has a huge impact over time with very little effort.

- You are responsible for everything in your life. This means how you position yourself, what choices you make, and how you react to the world around you. Of course, you're not responsible for a lightning bolt that drops a tree on you, but you are responsible for the ways you choose to react. You are responsible for being out in the storm. Accept the responsibility to make your life what you want it to be.

- Keep moving. This rule covers positioning yourself to avoid problems and take advantage of and/or create opportunities. It also covers continuous learning. The same goes for physical activity. We're not built to plant ourselves like trees. You can always be moving to improve your position.

- Be prepared. I learned this as a Boy Scout and it's such a cliche that most people ignore it altogether. This goes back to taking responsibility for your results. Starting a new job? Learn everything about doing it well. Continuous learning is a must in our competitive world. Getting married? Know your spouse and yourself. Have those tough discussions about money, kids, and family values beforehand. Stay healthy.

"Illustrations by Dall-E."

- Be good. Build a world where life gets better for everyone. Work toward it. Justice, wealth, and resources are still not evenly distributed. Evil exists and bad things happen but the good you can put into the world is what makes life worth living. Work toward a better tomorrow.

- Be lazy. Discover what's truly important to you. Make sure you're doing the right things before you spend effort doing them efficiently. Focus on being effective first and then on being efficient. Only do worthwhile things. Do them as well as possible and as efficiently as possible. Effort spent anywhere else is just wasted.

- Gung Ho. This is something the Marine Corps taught me that is deceptively powerful. The literal translation is "working together". No individual can be as effective as a group of people united in a common purpose. The family is stronger than the individual. The community is stronger than the family. If we can ever unite all of humanity to work toward a world that is objectively better for all of us, imagine how effective we could be. By the way, this also calls back to the earlier guideline about being lazy. Nothing is more effective than all of us working together for a common purpose.

"Illustrations by Dall-E."

"To travel fast, go alone. To travel far, go together"
— **African proverb**

Benefits:

Better outcomes — No guarantees, but they have been working for me. They improve my odds of making good picks by avoiding bad options and focusing on important things. They help to continually move me toward better positions. You'll need to choose the rules that work best for you.

Less worry and clutter — By using these as a starting point, I don't have to worry that I'm not making the "optimal" choice. My choice will at least be a good one. Don't waste energy or worry about things that are completely out of my control.

Simpler and quicker decisions — For the things I can influence, these rules give me a much simpler and quicker way to evaluate things. If something passes these guidelines, I can be pretty sure it's not going to harm me or others.

That's My Perspective...

4-29-24
Learning how to learn

Get MUCH better at learning, even when you think it's already your superpower...

"Illustrations by Dall-E."

I was always one of the bright kids in school. Learning for me seemed easy **until it just wasn't easy anymore.** In college, I hit the wall on several fronts. My natural unwillingness to study after class conspired with difficult material to learn, and no one to spoon-feed me knowledge. That put my problems with learning into sharp focus.

It didn't occur to me that **I had never learned how to learn.** I blamed holes in my math skills on a high school Algebra teacher I didn't like. I had no clue how to learn from large-scale lecture classes. I had forgotten all the things I learned "just for the test". The only study I did was a cram session before a test. My problems with learning were someone else's fault or just bad circumstances. I wasn't failing exactly... but I wasn't learning either.

I decided enough was enough and joined the Marine Corps. That is one of the best decisions I ever made, for a LOT of reasons. An important one was learning that **I had to take full responsibility for my own learning.**

I did learn there because they taught me self-discipline. At the time, the Vietnam War was raging, and the Marines needed people who could repair war-damaged electronics. They also had a **very good incentive program**. Five days a week you would study. On Saturday, you'd test on the previous week's studies and your cumulative knowledge. If you passed, you remained in school for the following week. If you failed, you went to Vietnam.

"Illustrations by Dall-E."

I could learn complex and difficult things, but it was a ton of hard work. I learned electronics repair so well that they decided to send me to cryptographic repair school. I became valuable enough that the Marine Corps sent me to Japan and didn't risk my one-and-only butt in Vietnam. I had to develop those skills by grinding away as if my life depended on it.

There are much better ways to learn. I eventually built my career by being the guy who would take on any job and learn how to do it fast. I am a committed lifelong learner. The difference for me is that **now it's much easier for me to absorb and deeply understand complex materials**.

The ONE skill missing throughout my schooling was the skill of learning the best ways to learn (backed up by science and not just the theory of the day).

My approach to learning:

1. Our attention span is limited to about 20 -30 minutes so **break up study sessions.**

2 **Get an overall basic picture** of the thing you're trying to learn. This gives you a backbone to tie the details together conceptually. We understand things better when we can relate them to things we already know.

3 **You remember things best when you have to work to fully recall them**. The harder you have to work, the longer you retain the information. Do something unrelated for a while and then return to the original subject, forcing yourself to recall what was discussed.

4 **Spaced repetition,** repeating the information at different times and over increasing intervals, helps build long-term skills. I use the **Readwise** app to save my notes from books I've read. It automates my review process by

creating questions and review cards from those notes and emailing me a selection daily.

5 Testing yourself helps build skills. Pro tip: When you are reviewing a subject or making notes about it, make up some test questions or problems and put them in Readwise or on a flashcard app such as Anki.

6 **Get immediate feedback** about your answers and keep the questions in your reviews until you have mastered the material — then review a few times later on to make it stick.

7 Pay close attention as the material is presented and then immediately after a class, review the material and summarize the key points in your notes. The work of recalling the lesson and summarizing it as if you were going to explain it to someone else helps fix the memory.

8 **We must sleep to consolidate long-term memory**. **Sleep, or even just a short rest, is one of our best learning tools**. The resting brain "consolidates" the memories Just as your muscles need sleep to repair, your brain needs sleep to build the physical connections between neurons that form memories.

9 Pro tip: Mentally review your subject just before bed. You'll remember more.

Teach someone else. As you get an understanding of a new skill, demonstrate it just as you'd do if you were teaching. This helps firmly set the memory skill and also helps you identify weaknesses.

LEARNING IS A LIFETIME SPORT

Resources:

"Learning How to Learn": Oakley, Barbara, Terrence

Sejnowski, and Alistair McConville. 2018. Learning How to Learn: How to Succeed in School Without Spending All Your Time Studying; A Guide for Kids and Teens. New York: Penguin.

"A Mind for Numbers": Oakley, Barbara A. 2014. A Mind for Numbers: How to Excel at Math and Science (Even If You Flunked Algebra). New York: Jeremy P. Tarcher/Penguin.

Khan Academy and Khanmigo — Free online classes on almost any subject. Khanmigo is an AI tool specifically developed as a tutor. I'm currently using Khan Academy classes in Algebra to fill in my knowledge gaps and strangely enough, it's more fun than Sudoku. Who knew?

Coursera — Excellent online tools. Classes from all over the world done extremely well. Take the *"Learning How To Learn"* online class here for free.

Farnham Street — Why reinvent the wheel? This blog and books like *"The Great Mental Models"* form a wonderful framework for learning.

"The Great Mental Models Volume 1: General Thinking Concepts": Parrish, Shane, and Rhiannon Beaubien. 2018. *The Great Mental Models Volume 1: General Thinking Concepts.* Latticework Publishing.

Readwise: A platform designed to help you get the most out of your reading by organizing and reviewing your eBook and article highlights using spaced repetition.

Anki: A free and open-source flashcard program that utilizes active recall testing and spaced repetition, drawn from cognitive science, to aid in memorization.

That's My Perspective...

05-06-24
The "M" word...

The Maintenance Manifesto:
Thriving, Not Just Surviving

"Illustrations by Dall-E."

Entropy:

As time passes, entropy increases. Imagine entropy as the chaos or randomness within a system. When things become more disordered (hot drinks cool off, bridge cables rust, and your body gets weaker), entropy increases. You must add energy to the system to prevent entropy from completing its work.

Maintenance:

Maintenance is the way we inject extra energy back into the system, whether it's reheating a drink, painting to fight rust, or working out to stay strong. We downplay the importance of maintenance in our thinking. The phrase "routine maintenance" captures the flavor of how we see it. A necessary evil. That's the wrong perspective.

A Maintenance Manifesto:

Rule 1. Without regular careful maintenance, EVERYTHING deteriorates.

Maintenance is the work required to KEEP what we've already got.

Keep maintenance top of mind. Think of our progress since the beginning of time, from agriculture, indoor plumbing, and sanitation to modern genetically engineered vaccines, science, AI, and quantum computing.

Every, Single, Thing, needs maintenance. Institutions like laws, money, and even democracy itself need regular, careful maintenance and repair. Infrastructure degrades over time. Technology becomes obsolete.

"Another flaw in the human character is that everybody wants to build, and nobody wants to do maintenance." — Kurt Vonnegut

Rule 2. Make maintenance an important and integral part of everything you do.

It's natural for us to strive for the next new thing. By all means, keep your eyes on the prize but remember that the surface you stand on can fall out from under you if you haven't done the maintenance.

Pro Tip: Put maintenance on the calendar as your second block of time every day. Productivity books recommend doing your hardest/most important task first every day while you are at your best energy level. That's a great approach… Then attend to the **important but not urgent — maintenance.** Here are some examples:

- At work:
- Improve your workflow.
- Make sure you've kept your commitments.
- Clear the inbox.
- Find ways to streamline the bureaucracy.
- At home:
- Pay the bills.
- Declutter.
- Finish repairs.
- Exercise.
- In the community:
- Stay informed and vote.
- Maintain your network of friends.

"Only floss the teeth you want to keep." — Marine Corps Training Film circa 1971

Body/Mind

Your body and mind BOTH need exercise, nutrition, and sleep. Think of them like a building code (the absolute minimum requirements you can get away with without the building falling or some other major safety hazard). They are not a luxury add-on, but the minimum. Without food and exercise, you can't build muscle or learn.

Without sleep your body can't repair itself or create long-term memories.

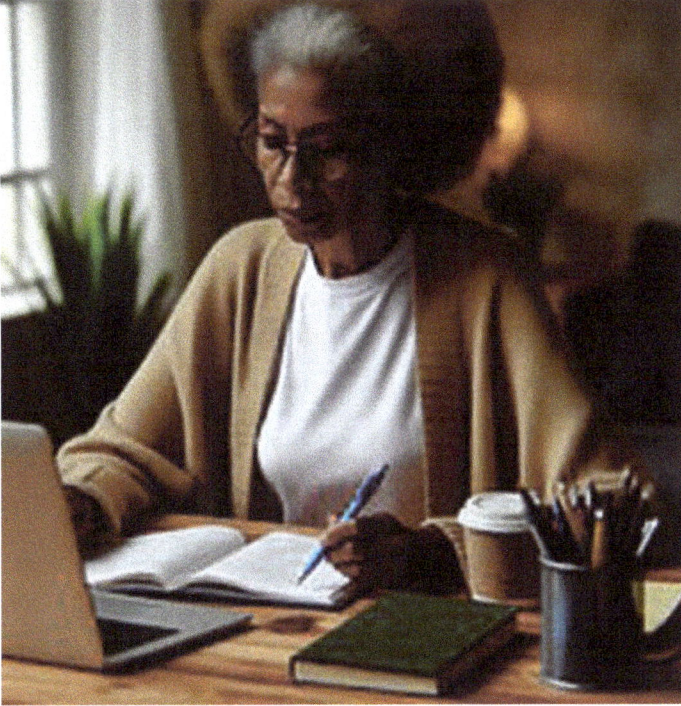

"Illustrations by Dall-E."

Rule 3: Make maintenance into a ritual:

"Make Your Bed" — Admiral William H. *McRaven*

Use care while doing the small things, like making your bed, flossing your teeth, cleaning the food prep area, doing repairs, etc. Create a set of routines that become rituals and then, habits. Morning routines are usually the same every day, but your Wednesday routine might add vacuuming. Getting your hair cut might be a Thursday's add-on every few weeks. Treating each with the care of a ritual helps to focus you for the day and make certain you come home to a well-made bed and clean house. This positions you better for every following day.

"Illustrations by Dall-E."

You can find real joy and meaning in the everyday work needed just to keep going.

"Before enlightenment, chop wood, carry water. After enlightenment, chop wood, carry water. — Zen saying

That's My Perspective...

05-13-24
Pet the cat

Getting real with retirement...

Since May is mental health awareness month, I thought this would be a good time for this post. -Tony Pray

For the first year after my retirement, I was a little batty. My poor wife dealt with my non-stop agitation, decades of relentless work habits, and compulsion to always be busy with saintly patience and the occasional reminder that I was "supposed" to be retired. I mastered the craft of being a pest to Donna as she did her best to keep me on the rails. Fortunately, she's a retired nurse with decades of experience dealing with difficult patients. I just didn't realize I was one!

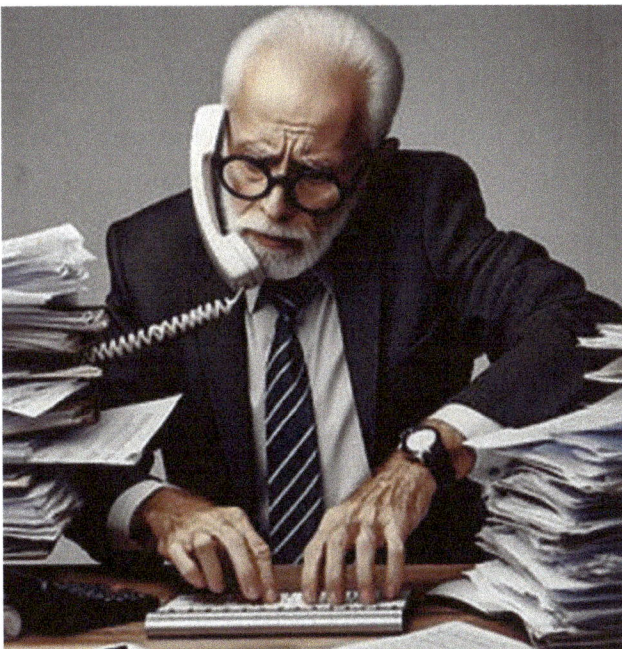

"Illustrations by Dall-E."

I had to be busy all the time and was getting very anxious.
Aikido. Exercise. Guitar practice. Write. Publish my book.
Learn to code. Build a website. Blog. Each activity, once a
normal part of my everyday life, became a box to check.
If I wasn't busy, I became stressed.

Slowing down to walk the dog, pet the cat, hug my wife, or
just read a book just for fun became chores to get over
with as soon as possible. Check the box and move on.

**Looking back on my first year of retirement, I accomplished a
lot.**

- **Exercise** — check
- **Book published** — check
- **Website built** and blog started — check
- Learn basic **Python programming** — check

I was proud of those accomplishments. but they kept me too
busy. I had to "downsize" my ambitions.

I needed help.

"If you dig a hole and it's in the wrong place, digging it
deeper isn't going to help." – Seymour Chwast

Donna suggested that I could research ways to find my own
helpful solutions. There are tons of helpful books online.
To help with sorting through the choices, I built an AI
prompt looking for the most useful tools.

I could do more for my mental health without a professional
than I realized. I eventually connected with a mental
health practitioner who agreed my approach had been
working well.

1. **It starts with getting enough sleep**. This has never been
a problem for me. I can sleep easily pretty much

anywhere. I added an afternoon nap to the routine just to be sure I was getting enough. It helped more than I thought possible.

2. **Get enough of the right kind of exercise.** My weekly high-intensity Aikido workouts were doing more damage than good with torn tendons, neck problems, and muscle pain to prove it. Subtracting the "wrong" exercise and adding a simple 30-minute walk and stretch routine every day made all the difference.

3. **Be Present, especially for the small joys.** The help and happiness I can give to others matter more to me than any self-imposed goals. Petting the cat isn't just something to do to get her to stop pestering me, it's pure joy for her and calming for me when I slow down enough to enjoy it.

4. **Identify the things that trigger my anxiety.** Learn to interrupt the cycle that leads to spiraling anxiety.

   a. **Stimulus** (My triggers)

   b. **Response** (The way they make me feel)

   c. **Reward** (Find a response that calms you down, not one that spins you up.)

5. **Medication can help.** Get professional help if any problems you're having are severe. For me, it meant getting a telehealth visit with a counselor to confirm that the things I'd started doing on my own were doing the job.

I gradually adjusted my days to allow for time with my wife and friends, quiet time, and time to simply be myself, pet the cat and enjoy the day.

Now I've simplified things even more. Each day I set aside a couple of hours for "work" (blogging or writing or research), "maintenance" (making the bed, exercising, cooking breakfast, and cleaning up,) and "live".

The whole point of life for me is to be contributing (this post for example) and enjoying time with my family, friends, and, of course, pets.

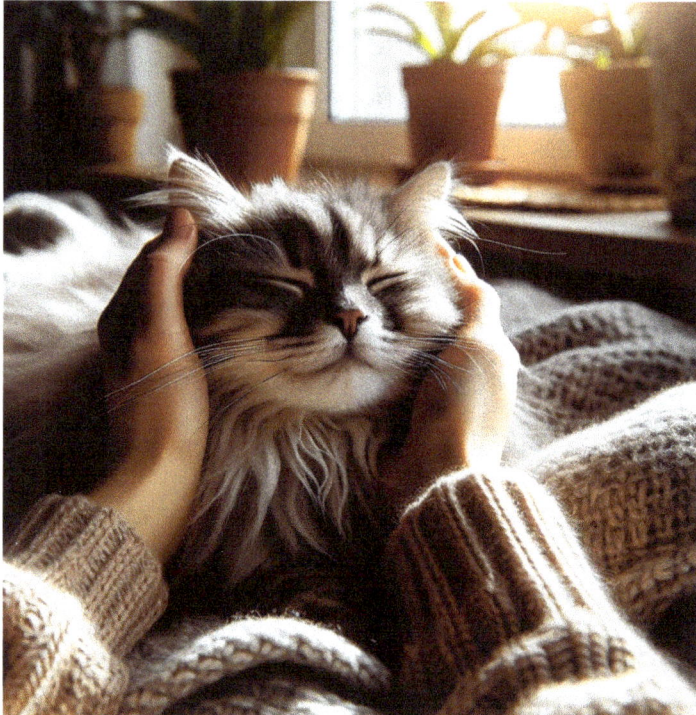

"Illustrations by Dall-E."

That's My Perspective...

05-20-24
Is It Ethical?

What to do, what to do, what to do?...

Ethics is a **set of moral principles** or a **theory or system of moral values**. It is a way of dealing with what is good and bad.

It's hard to pin down what's ethical in the "gray" areas. Especially when you're under peer pressure (Everyone does it) or in a hurry (Make up your mind!). It's far too easy to accept a customer's offer of free tickets, flirt with a coworker, or expense a meal that really shouldn't be. I'm not talking about the big and obvious ethical issues (obviously you shouldn't kill anyone) but the small, everyday choices you have to make.

"Illustrations by Dall-E."

It takes too much time and effort to deliberate about everything as it comes up. You need to be able to live your life with integrity without a lot of delay.

The simple solution to the problem of dealing with ethical questions is to build your own code of ethics.

A code of ethics is a set of **pre-existing rules** that you've already thought over and committed yourself to follow. It can save you and your reputation.

Mapping your own ethics code:

Define your values. Is it ever okay to take or accept something you haven't earned? Is it unethical to eat meat? How about spanking a child? Causing pollution? Smoking some weed? Here's my broad general definition that can serve as a basis for specific decisions.

"Illustrations by Dall-E."

Does it cause unnecessary harm or suffering to others? Also, consider legality and the rules of conduct we normally expect from others. Expensing a meal improperly might not be illegal, but it can harm you and your reputation.

Does it pass the "sniff" test? Act as if this decision would be reported by a very unsympathetic but honest person to your family, friends, and employer. If there's even a whiff of impropriety, it can harm you.

"Caesar's wife must be above suspicion."
— Julius Caesar

Sometimes harm to one being is necessary to help another. We all must eat to survive. As a society we have decided it is okay morally to eat meat. However, it is immoral to cause the animals unnecessary suffering. **You must weigh the costs carefully.**

Consider the costs that are put on others. Is it ethical to burn coal, profit from the sale and mining of the coal, and leave the "external" costs such as carbon emissions, mining tailings, lung diseases, etc. to be borne by other people? Go deeper and weigh the consequences of NOT burning coal in today's world.

Position yourself so your default choices will be to do what is best for you (and others) in the long term.

What kinds of ethical issues could affect you?

- Accepting or offering bribes or "incentives".
- Abuse from or by managers or others.
- Sexual misconduct including harassment.
- Use of force or coercion.
- Bullying.

- Cyberstalking.
- Lying, cheating, and stealing.

How will you commit yourself to dealing with them?

Why is it important?

Lay out your personal code of ethics before it's needed so you can act decisively in a way that's good for your long-term interests.

It's simpler to defend your choice. Saying "I've made it a rule to avoid things that can backfire" is something that most people won't challenge.

You will not be caught flat-footed when a quick decision with moral implications is needed.

Less day-to-day stress because this can be a "Once and Done" exercise. Instead of the slow and effortful process of reviewing all the pros and cons of a choice, you can skip to the end for many of them.

You'll also have a good starting basis if something totally unexpected happens.

That's My Perspective...

05-27-24
Earned Value

Clear feedback for honest work...

One problem with managing a complex project is getting an accurate sense of the progress and how much (extra) time and money (yours) it will take. Every project faces scheduling issues, people getting sick, delays with materials, etc. For example: in a home remodel project, when you ask your contractor "How's the work going?" you may receive feedback like "We're on track". However, you need to know the details about what's happening. That's where "Earned Value" comes in.

Earned value helps to create clear, reliable two-way communication and prevent little problems from exploding.

How Earned Value Works:

I'll use a home remodel project as an example, though it could be any complex project. **Complex projects have dependencies** that affect time and costs. For example, the foundation can't be dug until the building permit is issued and the permit won't be issued until the blueprints are complete. Each item in a dependency chain affects anything following it.

Project managers assign *milestones* to the key tasks in the chain of tasks from start to finish. They can also assign a dollar or percentage value (for the release of funds to the contractor) to the completion of that task. Here's a simple example:

"Illustrations by Dall-E."

Remodel Milestones

Milestone #1: Design completion and permits approved- 10%. — March 1

Milestone #2: Materials delivered to site — 30%. — April 1

Milestone #3: Demolition and cleanup — 5%. April 3

Milestone #4: Framing and electrical inspections passed — 25%. — May 15

Milestone #5: Final punch list inspection accepted — 30%. — May 24

In this example, each milestone represents a block of **Earned Value.** Passing a milestone means that the work is completed by the contractor **and accepted by you (or your designated inspector).** Passing a milestone means that the value for the work leading up to it has been earned and can be paid out.

Milestones can also be assigned a due date. This allows you to know if you are facing delays, deal with issues before they get too far advanced, and hold the contractor accountable for the progress. The earlier you detect problems, the faster and cheaper it is to deal with or avoid them.

Why it matters:

- **Better communication**. Instead of getting the answer "It's all good" when you ask how things are going, you can expect "We're finished with design and permits and expect the materials onsite by next Monday."

- Better understanding of the costs, schedule, and issues affecting the project.

- **The contractor has incentives to properly complete each stage of the work.** Even if it turns out it's a bad contractor you must fire, you'll know it much sooner and that helps limit the damage.

That's My Perspective...

06-03-24
Make Better Guesses

Veering a little closer to reality...

Steering closer to our course

Bad guesses are the silent saboteurs of our future. When it comes to life's big questions—like how much to save for retirement or how long a project will take—guesses can be costly.

Base rate approximations give us a reality check. Don't guess how long your project will take until you know the typical duration for someone with your skill level. For instance, a pro plumber might replace a toilet in 10 minutes, while an average homeowner could take 2 hours.

Thinking in probability adds nuance to our decisions. It's not just about whether it'll rain, but how likely it is. That project might take 3 to 6 months, but you're 70% confident it'll be closer to 4. This way, you're prepared for the worst but hoping for the best.

Revising bets keeps you agile. New information? Change your estimates. It's like rerouting your drive home based on traffic updates.

Removing biases means questioning your gut. Love that stock? Double-check why. We often overestimate our favorites and underestimate the rest. **Negative thinking** can be a powerful tool here. Imagine the project is finished and recount every obstacle you overcame. This mental exercise forces you to consider potential challenges and alternative approaches.

Fermi estimation helps when you need a quick ballpark figure. How much does an ostrich weigh? More than a man, less than a car. So, somewhere between 150 and 3000 pounds. Now you've got a range to work with.

Bayesian reasoning is about updating your guesses with new evidence. Think of it as adjusting your bet as the race goes on, not just at the start.

The payoff? You'll make smarter decisions, save money, and avoid the stress of those "I didn't see that coming" moments.

Sharpen your guessing game. Use these strategies, and you'll find yourself making more informed, reality-based decisions. And that's a bet worth taking.

That's My Perspective...

06-10-24
Synergy in Writing: The C.O.D.E. System Meets AI

Working with AI through the C.O.D.E. Method

Hey there! This post is brought to you by a human and an AI (Copilot), teaming up to share the C.O.D.E. system. Together, we're mixing human touch with AI smarts to give you the best of both worlds.

"Illustrations by Dall-E."

I'm drowning in the deluge of data. The volume is staggering! The flood is relentless, from podcasts to books to emails to articles to videos. Navigating this mess, attempting to retrieve elusive thoughts, leads to lost insights and hours piled onto each project, and the certain knowledge that I'll never catch up. This isn't just a personal struggle; it's a widespread challenge everyone faces

To navigate the ocean of ideas, the C.O.D.E. system serves as my compass. C.O.D.E. is a process designed by Tiago Forte to enhance personal productivity and creativity. It stands for **Capture**, **Organize**, **Distill**, and **Express**.

- **Capture:** Like casting a wide net, we gather every idea of interest or potential value into a trusted digital capture system, ensuring no idea is lost. Tools like Evernote, OneNote, or Notion, with their cloud backup and synchronization capabilities, are ideal for this purpose. It's important to capture entire ideas as separate notes, which can later be remixed and repurposed.

"Illustrations by Dall-E."

- **Organize:** We chart our course, categorizing the captured content into a structured system for easy navigation and retrieval. Methods like PARA (Projects, Areas, Resources, Archives) or simply using folders and tags can turn a pile of notes into an organized library of information.

- **Distill:** As we refine each idea to its most valuable form, distilling the essence, it's our unique perspective and knowledge that determine what's worth expressing. This step is about making connections between different ideas and synthesizing new insights.

"Illustrations by Dall-E."

- **Express:** Sharing our distilled insights means making waves with our creative and productive output. Starting with an outline or narrative framework helps me do a better job. Try starting with storytelling techniques, since we're wired for stories, not so much for lists and data.

C.O.D.E. puts speed, purpose, and precision into my work, transforming the overwhelming tide of information into actionable ideas. It's about working smarter, not harder, and helping ideas flow

That's My Perspective...

06-17-24
AI Withdrawal

I'm not quite a cyborg, at least not yet...

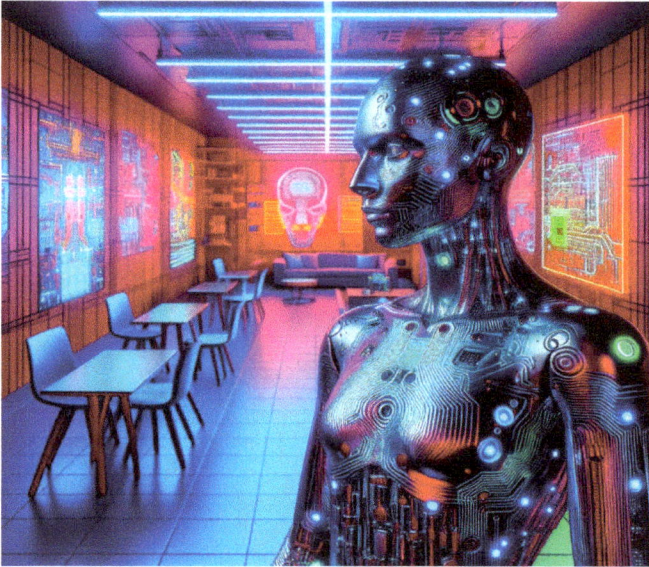

"Illustrations by Dall-E."

AI hasn't quite reached the promised land yet, but it's danger close. In my last post, I experimented with using AI (Copilot in this case, using ChatGPT as the underlying AI) **as my full-fledged writing partner.** We looped through each step of producing the post. We even argued (respectfully) about my tagline at the end of the post. I have experience using AI for routine tasks, but this session surprised and worried me.

Prep work for this took much longer than just sitting down and writing the post on my own. However, it's a one-time task that might save work and create better writing later. (If you're interested, I've posted my "Master Preferences" prompt in Substack's Notes section.)

One serious problem with any current large language model AI is that it "hallucinates". It doesn't truly understand what it's writing. It has a sophisticated statistical map of the word most likely to follow the present one in the current context. It can be wildly wrong and still sound completely plausible and even logical. Here are some examples:

- I asked the AI to give me detailed instructions about the best way to index and tag several years' worth of notes. It provided me with very detailed step-by-step instructions to export, convert formats, store the files in a cloud server, etc. Several hours of work later, I reached the final step in this tortuous process — Mail the link for my OneDrive folder to the AI... (Think about this step for a second.)

- **It's an AI! It doesn't HAVE an email address!** When I pointed this out to the AI, it apologized. The AI doesn't understand an apology, but the response sounds quite good.

- I'm taking an online Algebra class for fun using Khan Academy (online tutoring). I was happy about a difficult problem I had gotten correct but wanted to review the solution step-by-step using their AI tutor, Khanmigo. When I worked on the problem with the AI, **it got the solutions to both sides of the equation wrong**. It doesn't have the answer key available, **and it did the math incorrectly**. The AI did apologize and acknowledge its error, but that mistake undermined my faith in the tutor's ability to teach.

None of these things would be important if the AI was simply bad at what it does. I know that Copilot will often create weird images. For a person, it can create someone with three arms. For a man tossing a life ring, it shows the man standing on top of the water. To show a group of Marines getting a Humvee out of mud, it shows people pushing hard from both ends of the vehicle. **I know to look out for those things.**

The real problem is that it does a fine job most of the time. Need some story ideas? Easy. Want to process a bunch of material and reformat it? No problem. Searching for a particular phrase? Cake. Want a suggested story outline? Simple. Want to do an index of thousands of notes? Easy-Peasy… until you realize you've blown 4 hours following hallucinated instructions.

It's like having a self-driving car that only glitches once in 10,000 miles. By the time it happens, you're not closely monitoring the vehicle, you're not asleep but you're not expecting a catastrophe either. It takes you some time to react and, even if you're wide awake and paying some attention, it's going to create problems.

I'm already a cyborg by some definitions. I have hearing aids, glasses and artificial knees. I use digital technology extensively as my "Second Brain" and external memory. I'm an early adopter for new technology.

I'll continue experimenting and learning with AI firmly in the mix. Why? Because it will so much better. Today's AI is the worst version we'll ever work with, and it can be stunning. Some of my posts will be written by me and the AI working together. Some of them, like this one, will rely on my analog skills.

That's My Perspective…

06-24-24
Enough?

What is "enough" money?

As a kid, I believed that once I made as much money as my dad made, I'd have "enough money." At the time, he made about $10,000 a year in his machinist job...

He raised our family with three children on a single income by working a lot of overtime, odd jobs, and managing rental apartments on the side.

I entered the Marine Corps and learned I didn't need much money. I was happy just doing my job and learning new skills (their electronics school led to my eventual choice of career). Still, I enjoyed "stuff" like scuba gear and racing bikes. I was always a little in debt and chasing "more money."

Daydreams

After leaving the service and starting my own family, having children led me to understand that the "toys" were not important. I sold the scuba gear, underwater camera, bike, trailer, and all the other toys. "Enough" money changed drastically, and I took a series of jobs to meet my family's needs.

I wanted "enough" to quit the jobs and find a real career.

By the time I found that career, I had remarried into a new, blended family. Better salary and benefits followed.

That's when I realized I'd never have "enough money" to compensate for lost family time or retire with complete financial security. I'd never have "enough" until I figured out for myself what I'd need to have the life I wanted.

"Illustrations by Dall-E."

I believed that I needed the security of "enough money" but it slowly dawned on me that I had been a very happy kid, even though when I looked back, we were very poor. I'd been happy in the Marine Corps, even though my pay as a sergeant wasn't going to allow much luxury. **Eventually, I realized that "enough" was about more than money.**

I spoke with my wife, did some soul-searching with her, and realized a simpler life would be better. We sold our big house and downsized to an apartment. Sold the furniture from our former five-bedroom place and started the process of minimalizing our "stuff". **We came to realize that we had enough!**

Turns out that we had "enough money" to live a modest, but NOT a poor life. I've been retired for five years, and money can be a worry at times but there's no way I'd exchange more money for the "enough" that we've found.

Enough is enough.

That's My Perspective...

07-01-24
Energy Management

Getting the best mileage out of yourself...

Energy management is key to boosting your productivity. Productivity and self-improvement books focus on time management and goal setting. Goals, priorities, and time management are important to getting things done. However, **having prioritized goals tracked in your calendar gets you nowhere unless you have the energy and will to do the work.**

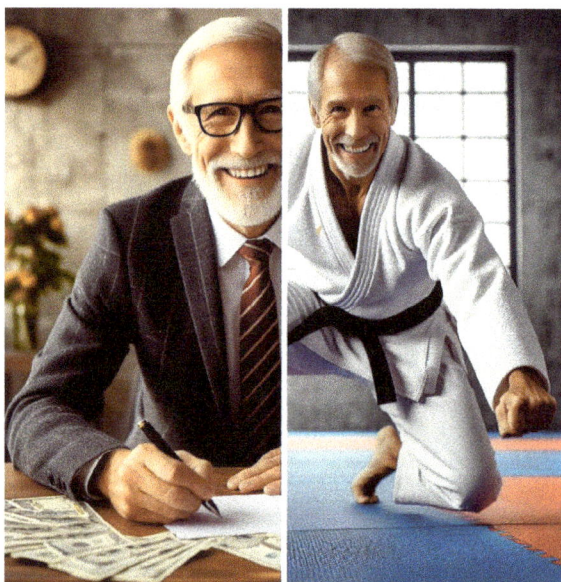

"Illustrations by Dall-E."

Writing checks the body can't cash...

I love Aikido. The challenge never gets old... but I did. Just a few months ago, I retired from my Aikido practice. I'd been telling myself that I could still do anything the "kids" could do, it just took me longer to recover. So, I limited myself to one kid's (8 to 12) class and one light practice adult class per week.

At 72, a torn tendon forced me to face the fact that even though I still have lots to give, my body limits me. I don't have the stamina or recovery capacity.

I've also learned to **pay attention to what my body is telling me** and stop writing checks my body can't cash.

Energy at work

As my career advanced, I gained the freedom to have my work judged by the projects I delivered and not by how "busy" I was or how many hours I put in at the office. Once results mattered more than appearances, I was free to succeed or fail based on my own efforts. **That let me manage my personal energy budget and focus my efforts to be effective.**

"Illustrations by Dall-E."

Here are some of the things I learned that I'd like to share:

Make sure that you're digging in the right hole before you try to get more efficient. Working hard on the wrong things is worse than doing nothing at all.

Wrong hole…

Most people's energy is highest at the start of the day. Pick the most important thing you can be doing right now, the one that will make your life easier when it's done and is probably the hardest and most unpleasant. Do it first with a serious focus. Rinse and repeat every day.

Sprint or jog? If you determine that your current task is something that can be dealt with in one or two concentrated sessions, go into sprint mode. This is fine for a one-off or for dealing with an emergency. But if you have too much to deal with in a short sprint, you need to slow down and settle in for some LSD — Long, Slow Distance.

Think of it this way, just one page a day would produce a 365-page book in a year. One brick a day will build a substantial wall. **Make time into your friend and match the work to your energy level throughout the day.** When new things come in prioritize them and get them onto your schedule while keeping your own energy budget in mind.

The Legs of the Energy Management Triad:

1. **Prioritized Goal** (aligned with your purpose). Make certain you're doing the work that will be most *effective* rather than the most efficient. Dig in the right place and only then worry about being efficient.

2. **Time management**. Your time is your most important non-renewable resource. Make it count for more. Do the most *important and impactful* work first.

"Illustrations by Dall-E."

3. **Energy management. Know what works best for you.** I'm a night person who starts slowly but I can sprint if needed. By making it a practice to do the primary task first, I have the energy and the will. Handling the routine stuff later in the day lets me jog or even walk so I can cover a lot more ground.

Resources:

Allen, David. Getting Things Done: The Art of Stress-Free Productivity. Penguin Books, 2001

Keller, Gary, and Jay Papasan. The One Thing: The Surprisingly Simple Truth Behind Extraordinary Results. Bard Press, 20132

That's My Perspective...

07-08-24
Penny's Tale

Lessons from a curious beagle...

About twenty years ago, Donna and I rescued a beagle named Penny. After getting to know a neighbor's beagle, I wanted my own. Penny faced abandonment as her owners moved away. Our real estate agent heard about this, and we suddenly had a new family member.

"Illustrations by Dall-E."

Before Penny, I had settled into a "life of least resistance". Like water, I flowed to the easy chair after work. The TV, my phone, and a snack became my buddies. Screens, notifications, doom-scrolling, --- slowly turning into a living room puddle. I thought a dog might help me get off the couch a little more…

Always be moving. Penny had plans too.

Penny had a huge helping of curiosity, especially for new smells. We had a favorite place where Penny could get off leash, run and play, and chase rabbits until she was worn out. Our routine was to circle the big electrical substation near us. They built a large berm around the plant, and it butted against woods on two sides with a creek on the third. Lots of cover for rabbits, hills to climb, and trees for shade. It was beagle heaven. The walks became my daily R and R.

Be tenacious and focused. When she caught a scent, Penny would let out a baying howl that I could hear from half a mile away. She'd run full speed with her nose just an inch above the ground, sometimes swapping her head end for her tail as the rabbit trail doubled back. The hillside was steep, and she'd sometimes lose traction and sideslip down the hill before getting back on the scent. As she got close to the rabbit, it would go for cover in the heavy underbrush. The white tip of Penny's tail wagging at a furious rate was the only thing I could see of her unless she jumped up to get a view over the grasses. The rabbit usually got away. She didn't seem to mind that, she'd come running back when I called her with a big grin on her face.

Sometimes, she didn't realize she was smelling trouble... Her beagle nose, that finely tuned instrument, led her into unexpected encounters—the kind that makes you question the sanity of a dog chasing scents.

Once, she bayed near the creek and went into a stand of trees along the bank. I lost sight of her but could hear her excited barking, so I followed her in. A few yards in, was a mud flat where I saw Penny (actually, just the white tip of her tail) waving like a signal flag in the tall grass.

Curiosity is great, but don't stick your nose where it doesn't belong. I stepped up to see that she had cornered the largest alligator-snapping turtle I'd ever seen! About 35 pounds of thoroughly pissed-off evil, with jaws like a bolt cutter. Penny snapped at the turtle, The turtle, unimpressed, popped its head back into its armored shell.

"Illustrations by Dall-E."

As Penny retreated for a second, the turtle's head snapped out with breathtaking speed and missed removing Penny's nose by a tiny fraction. But Penny wasn't one to back down. Before I knew it the scene repeated. Luckily, beagles come equipped with a handy-dandy beagle removal handle. I grabbed Penny by the tail and dragged her away, her eyes still fixed on the turtle. The turtle hissed at us like dinosaurs must have sounded, creepy.

A few minutes can make a day great! It usually took us about twenty minutes to circle the substation, but every day I learned from Penny a bit more about how to stay in the moment and love every minute.

Penny thought that anything with four legs was another dog. So, when a woman riding horseback came down the access road Penny started barking to protect her pack. Penny had never seen a horse before. Hackles up! The lady paused, her steed studiously ignoring the little beagle dancing and barking a few yards away. I saw an opportunity and asked her if it was all right to approach the horse with Penny. She said, "Sure, give it a try."

Have courage but don't get crazy about it. Horses, at least from Penny's reaction, are scary when you get up close. They get larger and, if you are a sweet-tempered beagle with nothing to do but protect the family, you react by getting frantic. The horse stayed perfectly calm, but we never got closer than about 20 feet from it before she started to freak out. She was full of courage, (Think what your reaction would be if you were confronted with King Kong) but she wasn't crazy enough to let that huge dog get too close.

Learn from the mistakes others have made and do the homework (whether it's for a pet, a home, raising a child, or understanding car maintenance). Penny once came in and sat by me in my home office. She gave me a sad look, as beagles do when they want food or attention. I was preoccupied and ignored her until she suddenly screamed! A short piercing scream as if I had stepped on her tail or something. Then she fell over on her side. She immediately tried to get up but fell over on her other side, whimpering now and in pain. I thought she was having a stroke!

After several anxious hours, we learned Penny had ear troubles. It's a common problem for long-eared dogs but I had never done any research about taking care of beagles. I was ignorant about beagle ears until the emergency vet told me her eardrum had been infected and the pressure made it burst. That threw off her balance so she couldn't stand.

Look at things your own way. One course of antibiotics later, she was back with us and almost as good as new except for one minor thing. From that day on, she carried her head tilted slightly to one side. This made her look as if she always had a question on the tip of her tongue.

Live in the moment and love it! Penny lived a long and happy life with us after that. I finally lost her several years ago. She was one of my favorite people in the world. She showed me what loving the moment looked like! Getting off the couch is only the start. Do something that brings you joy every day. Laziness isn't rest; it's stagnation. Unplug, reset, and reconnect—because life happens beyond screens.

That's My Perspective…

07-15-24
What's Wrong With "Best Practices"

Raise the Bar...

"Best Practice" is a harmful cliche. People use the phrase carelessly. Will it always work? Will it be "Best" in different circumstances?

For me, the "Best Practice" approach turned out wrong at least once. Working on an international banking IT project, I shut down an endless Zoom call about a reporting dashboard for the executives by citing "industry best practices" as my preferred approach. That got me labeled a "thought leader." The result wasn't really a great fit. We finally arrived at the right solution through experimentation and testing.

"Best practices" can refer to processes, materials, organizing, or other ways of dealing with things. Using the label "best practice" can result in doing "it" in a way that works, which makes some sense. After all, who would want to do something that was proven not to work? But don't let best practices be the end-all.

"Best practices" should be a starting point. They are a valuable way to look at a tried-and-true method. They are a necessary place to start, not the best place to end up.

To get better than "Best Practices", try this approach.

1. **Start with a clear picture of your current actual process or practice**. You must overcome the inertia of "the way we do things". **People will revert to the default mode** unless you make the changed process the only acceptable approach or (much better) make it much easier/faster/more convenient/hassle-free.

Once upon a time, my team was stuck using Excel spreadsheets to track projects. Switching from Excel to project management software saved hours every week for everyone, but getting the team to change was still difficult. The learning curve was daunting. An hour's tutoring was all it took to get us over the hump.

2. Ensure you are getting over the "Current Best Practice" bar.

Imagine a restaurant that prides itself on using the freshest ingredients. Reviewing industry standards could tell them they could get better quality by sourcing from local certified organic farms.

3. Make a list of your three most pressing problems. What still isn't great from the user's view?

For example, A. Employee turnover, B. Profit margin, C. Speed to market

4. Focus on the most pressing one and think of ways to close the gap between today's situation and the desired result. Let's assume employee turnover is the biggest problem for this exercise. To close the gap you could pay more, train people, offer incentives, etc.

5. Reassess the first problem, considering the approaches for your second and third most pressing problems. Make sure your solutions don't conflict with each other. For example, paying people more affects profit margin—keep these

interactions in mind. Find a balance.

6. Do the same for your other issues. Make sure that you end up with a list of actions to try that don't contradict each other.

7. Pilot some variations of the most promising actions, especially those that reinforce each other. Offering better training can reduce turnover and help increase the profit margin. Used in conjunction with your marketing plan, it could make your company stand out.

8. Roll out the successful practices as your new "Current Best".

   Rinse and repeat for the next business problems you find.

   Pilot this approach to fit your needs, making sure the changes don't have unintended consequences. Test the approaches in reality and move toward a more coherent overall strategy.

   Continuous improvement, tested in the real world, is the true "Best Practice."

   That's My Perspective...

07-22-24
The Grand Unified Theory

Bending The Universe Your Way...

The holy grail for physicists is the Grand Unified Theory.
It should explain everything from the largest scale
(explained by the theory of relativity) to the smallest scale
(covered by quantum mechanics). It should provide a
single coherent explanation for everything about the
physical properties of matter, spacetime, and even
gravity. It would unify all of physics in a single theoretical
framework.

My Grand Unified Theory focuses on bigger goals... how to
bend the universe to your will. (...evil grin)

"Illustrations by Dall-E."

Bending the universe...

Label the physical universe as level one. It's the matter and energy that build everything.

Level One: The Physical Universe

The matter and energy that build everything. This is everything we can sense or have ever sensed. Physical stuff. *Everything* is composed of energy, even the matter around us. Matter is made up of enormous amounts of energy.

Consider this: the energy in a milligram of a fingernail clipping has energy equivalent to **21.5 tons of TNT**. Energy fields interact. So, I'm not being lazy when I take a nap... I'm trying to keep from exploding.

Level Two: Navigating the Cosmic Dance Floor

Imagine the universe as a dance hall, where each step and move is a choice that leads us through the melody of existence. In this cosmic choreography, **being in the right place at the right moment is like finding the perfect rhythm in a dance.**

- *Choose your steps wisely and position yourself well.* It's not just about grabbing opportunities. Your starting place on the dance floor sets the stage for your next move.

- Where you start matters.

- Move to better starting positions.

Level Three: Mastering the Dance

Your skills are the steps you've mastered and the rhythms you've learned. Each new skill is like a new dance move, enhancing your grace and impact. The more you practice, the more fluid your movements become, turning each interaction into an opportunity. Your knowledge is the music that guides you; keep learning new tunes to enrich your dance.

- Skills are the steps you've mastered.
- Knowledge is the music that guides you.
- What you know matters. Keep learning.

Level Four: The Role of Random Chance (Luck)

Luck is the way the universe has of making complex systems act unpredictably. The entire universe is a "complex system." Small changes in *any* part of a complex system can result in extremely large changes elsewhere (this is called the "butterfly effect"). Our reactions to these unpredictable events shape our path.

- How you react to random chance matters. React wisely.

Level Five: Interconnected Influence

Our physical bodies, thoughts, emotions, and movements are all patterns of energy. Energy patterns interact. Each "thing" has some influence on everything around it. The effects may be tiny, such as the effect your body's gravity has on the path a ball on the pool table takes, but they accumulate.

Small changes, like a single person catching a new virus (COVID-19 anyone?), can radically change the world. We live within an interconnected network where chaos sometimes hits. Most of the time, things are stable, until they aren't.

- We are patterns of energy, influencing and interacting with the world around us.
- Even the smallest actions can have profound effects on the larger system.

We live in a universe where positioning, skill, and chance will all combine to have effects. The position we occupy depends on the position we were in the moment before. The skill we bring to bear depends on the investment we make to learn.

Level Six: You have a limited amount of *control* in the universe, but you can *influence* much more. The Marine Corps taught me this lesson, ***Gung Ho!*** It translates as ***"Working Together"***. It's the Marine motto and it's the way to evolve from an individual to a superorganism.

"Illustrations by Dall-E."

The pack is stronger than the wolf. Start with a single wolf: dangerous and a scary adversary. Scale up to a coordinated wolf pack: Far more dangerous and effective.

Alone, we survive. Together we thrive! Start with a single human: a tool-making predator, adapted to flourish in every climate on the planet; also, a dangerous and scary adversary. Scale up to a village, which is far more effective than the wolf pack. At this scale, we tamed the wolf and made it our ally.

Each level of scale results in more effectiveness. Scale up to a city-state, a nation, or a global network. At this level, we cure diseases, communicate around the world in an instant, explore the solar system, and affect every living thing on the planet.

"Illustrations by Dall-E."

Together, we are most effective.

Level 6 suggests something. You can't *control* the world. Random chance makes controlling it impossible. In this hyper-networked world where everyone can reach and affect everyone, you can *influence* it.

Your influence bends the universe. Make sure it's bending where it should.

That's My Perspective...

07-29-24
Whiplash

Changing sides or?

Sometimes I write to sort out thoughts and feelings on different topics. This post is that type. **Please comment if you like. I welcome your opinions and know I don't have all the answers.**

Watching George Floyd's murder on TV caused me to think about things I hadn't paid much attention to recently. Since no one in my family ever owned slaves (or much of anything at all) and I try not to be a racist, I was pretty comfortable with business as usual. That has changed for me.

I've always had mixed feelings about racism and social justice movements. I'm a white guy with no particular hatred for anyone else. I know I have unconscious biases like racism, sexism, and "other-ism". Everyone does. I try to keep those in mind as I navigate through life.

Blue lives matter. I've always been a strong supporter of the police. It's a tough job where you see people at their worst all day, every day.

Black lives matter. No one should be at a disadvantage because of skin color, and no one should be brutalized without consequences.

Both are true!

Every life matters! I hate what I see happening with Americans getting to such extremes that supporting Black Lives Matter automatically means you don't support the police. Caring about any liberal cause makes you a traitor to the right wing. Mentioning that you're a gun owner gets the cancel culture machine revving on the left. You can only prove your loyalty by being even more extreme. You are required to see everything in black and white.

I call BULLSHIT!

"Illustrations by Dall-E."

Life is not black and white. I'm trying hard to live mine as well as I can and understand the best ways to navigate it.

It's true. We are more comfortable with people who are similar to us. That doesn't mean it's right, just the way we're built. If we work at it, we can learn to get more comfortable with others, but being human, we're also lazy about things that take effort.

I support LGBTQ+ rights. I'm boringly straight and too old for it to matter. That doesn't mean I approve of everything gay people might do or that I'm comfortable with it. It just means that it truly is none of my damn business. Must be a liberal snowflake... right?

"Illustrations by Dall-E."

I support the Second Amendment. I'm a former Marine who enjoys target shooting. Strangely enough, having a gun seems to mean that I lean right. I don't mind that. I'm right-wing again.

I disagree with the right about regulating gun ownership though. It's just boneheaded stupidity to give someone a gun without proper training. Wow, I'm left-wing again, or is this just me holding two incompatible things in my head?

I support gun regulation. We license and insure cars. I think proving you're an adult, with no red flags like felonies, restraining orders, or serious mental health issues, and thorough training should be the minimum bar — making me a liberal traitor, even though I think every adult should start with the right to keep and bear arms. It seems crazier than anything that a felon can be president but not legally own a gun. Hey, if you can't be trusted not to commit serious crimes, let's just give you command of the armed forces.

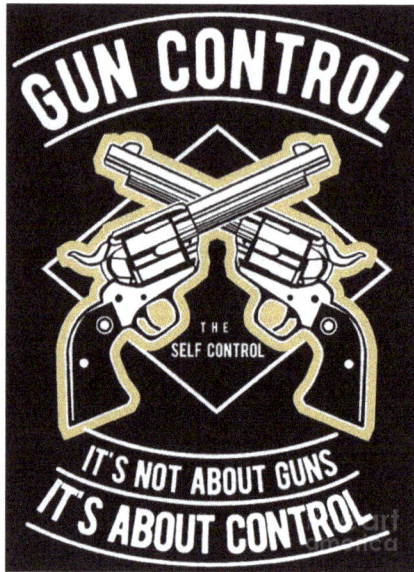

"Illustrations by Dall-E."

I support women's rights because wasting anyone's ability is stupid. (Holding people back because of their sex strikes me as a perverted form of stupidity.) Isn't that the liberal side's viewpoint?

Wait, isn't a smaller and less intrusive government a conservative thing? I'm getting confused.

"Illustrations by Dall-E."

Less government is better. The general statement is true, but that makes me a right-wing crazy… right? What I mean is that I'd prefer a much lighter touch and more effective government for the things that *really* require governing. We need laws to prevent abuse, not laws to enforce someone's view of what our morals should be.

The death penalty is a mistake. I don't think it's immoral to kill a killer to protect society. In a perfect world, we'd be guaranteed to get the right person every time. This is not a perfect world. The death penalty is wrong because innocent die people when we make mistakes in identifying and trying suspects. The death penalty removes the chance to correct our mistakes.

Hey, if someone is attacking my family, I'd have no problem personally administering the death penalty to save them. Our legal system's use of it assumes we're going to get it right *every time*. So, I'm on both sides of that particular issue at the same time. No problem, I contain multitudes.

Racial Justice: Everyone should be equal under the law. We need to keep working together to make that happen. No one gets to pick their parents or the circumstances they were born to. I believe the playing field should be as level as possible.

I support the rule of law (and I believe that letting anyone be above the law is a form of societal suicide.) At the same time, I know the law is not perfect. I believe laws should change as we learn.

Even if the other side wins, I'm not going to pick up a gun and go to war. (Except to administer the death penalty to anyone who harms my family… see above.)

It's hard to say what you mean anymore without getting heat from both sides! Political correctness went extreme and has been replaced by polarized insanity.

I'd welcome some sanity and your opinions, especially if you think I'm wrong. Who knows? You may have a point worth thinking about too.

That's My Perspective...

08-05-24
6 Lessons from the Mat

It's how I roll...

Special Thanks to Al, my ChatGPT AI assistant, for helping me make this post clearer and more engaging.

Every time I post, I ask myself "How can I make this post valuable for you?" Here's a glimpse into the process:

In the spirit of transparency and showing my work, I'm including a segment in this post you normally never see. I use a mental model to construct my posts called: **"Think Before You Ink"**.

Main Point: What is your main objective for this post? To show useful lessons I've taken from my Aikido practice that I can share with people who don't go in for martial arts.

"Illustrations by Dall-E."

Importance: Why should it matter to them? A genius learns from other people's efforts. (Bonus: It saves you a few bruises.)

Action: What action should this post inspire? I hope you'll give these concepts serious thought and make them part of your daily life.

Applying Aikido Principles to Daily Life

1. Awareness — If you can't get out of the way, the rest doesn't matter.

If you can't get out of the way...

1. **Situational Awareness**: It floors me when people wander through busy parking lots (often pushing a stroller with kids inside) with their attention firmly on their cellphones. Paying attention to what goes on around you is essential. You can't afford to be oblivious to the car coming your way, the bar you've wandered into, the next big layoff where you work, or the threats extremism of all stripes can cause. Evolution is still in effect. Even though we live in the safest and most prosperous times in history, **your results will vary**.

2. Harmony — Don't collide with your opponent, guide your opponent.

   - Someone is coming at you with a baseball bat. Block, or step to where your opponent can't hit you? In Aikido, as the bat swings, you step far inside the range where it can hurt you and get immediately next to your opponent. You then guide them to lose balance while protecting yourself from harm. You usually "assist" them to move much farther in the direction they were already going than they intended. Aikido is sometimes known as the gentle art of hitting people with the planet.

- Let's try it THIS way…

"Illustrations by Dall-E."

Understand others' perspectives so you can see things they might have missed and find ways to guide them where you want them to go. Use their goals to guide them toward mutually beneficial outcomes. Start from a position where you can be in harmony with your opponent's movements.

"I never allow myself to have an opinion on anything that I don't know the other side's argument better than they do." — Charlie Munger

3. Quick (And Effective) Decision-Making

- **Adaptability**: Aikido teaches you to make decisions on the fly. This skill is crucial in fast-paced environments where quick thinking is essential. If you're like me, you usually make urgent decisions by imagining a way forward and then giving it a quick check to see if there are obvious showstoppers. Not a great approach.

- When speed is essential, you don't have the luxury of time to brainstorm several solutions and then weigh the pros and cons of each approach, but you can get much better-quality decisions by adding two things (which take almost no extra time):

- **Consider at least 3 possibilities**. For example, "Should I fire this guy or not?" Becomes "Should I fire him, keep him, or transfer him to a job he's better suited for?" Almost any decision can be improved by considering another option

- Ask yourself "...**and then what?**" By doing this you force yourself to look at the consequences of the possible choices and how to deal with them.

4. Balance in Chaos

- **Staying Grounded**: Maintaining balance, both physically and mentally, even when everything around you is chaotic, helps in staying calm and focused under pressure.

- Be mentally prepared that **things will go wrong**. No one knows what will happen next but keeping a mindset that allows you to roll with the punches and be alert for opportunities will help keep you calm and effective. For example, when too many things are coming at you at work:

- Identify the most important action you can take (prioritize the incoming tasks).

- Do that.

- Repeat.

This approach helps me to stay calm whether I'm taking on multiple opponents or dealing with a lot of issues at work.

5. **Patience and Trust in the Process:** To get better at anything you need to realize **it takes time and a good process to learn complex new skills**. From your day-to-day perspective, you might think your progress has stalled. Try looking back six months and compare it with what you knew back then.

   The process you build should:

1. Give you a basic understanding of the skill.

2. Practice the skill in various ways.

3. Give you immediate feedback.

4. Loop until you master the skill.

5. Integrate the skill with the other skills you have.

   When I started learning to roll in Aikido, I could roll without hurting myself, but I thumped several times as I went over. A few months later, my rolls were silent, but I couldn't fall in a hard throw without pain. About a year later, I accidentally slipped on the icy stairwell at our community center and naturally went into a roll down 26 steps with no injury. My wife had seen me slip at the top of the stairs and thought I'd be dead, but when she ran up to see what had happened, I was standing at the bottom of the stairwell with a slow grin spreading on my face. Patience and trust in the process lead to mastering a skill over time.

6. Knowing the Difference Between Pain and Harm:

   In Aikido, you learn to differentiate between pain, which can be a part of growth and learning, and harm, which causes lasting damage. Aikido techniques such as joint locks can be excruciating without causing lasting harm.

Learning to deal with the immediate pain to reach a goal makes it possible to take calculated risks and push boundaries without causing long-term negative effects.

During training, applying a wrist lock can cause significant discomfort. This immediate pain does not result in harm. In contrast, ignoring your body's signals and pushing through harmful pain can lead to lasting injuries. Understanding the difference helps you take calculated risks and push your limits safely.

"Illustrations by Dall-E."

Unsolicited Endorsement of Readwise: If you're looking for a way to make the most out of your reading, I highly recommend giving Readwise a try. I've been using Readwise to help me retain and revisit important lessons from the books and articles I read for years. I can be confident that I won't forget that wonderful idea that I just found. I have all of my notes and highlights saved and I get a daily review to keep them fresh.

I don't have any affiliation with Readwise or get anything from promoting them. It's just **the best tool I've found to help make the lessons I work hard to find, stick in my mind.**

That's My Perspective…

08-12-24
The Most Dangerous Addiction

What Is It Costing You?

The most dangerous, insidious, and deceptively benign addiction I've ever faced isn't what you'd expect. It is a silent dream killer, an always-ready-to-strike snake in the grass, the spider at the center of an invisible web…

"Hi, I'm Tony, and I'm an addict."

My addiction? **Comfort.** It's not as obvious as some other vices but just as real. **The problem?** Comfort kills. It ruins lives, families, and communities. With no fuss or bother, it destroys possibilities and steals the best of what life could be. **My solution?** Leaning hard into discomfort isn't just about personal growth—it's about reclaiming our lives from complacency.

"Illustrations by Dall-E."

When I was about twelve, I remember something my Dad told me. "Son, you're the hardest working person I know." "You'll work twice as hard to avoid work as it would take you to just do the job." He had captured my attitude perfectly.

Comfort: The Silent Dream Killer...

I didn't mind hard work, as long as it wasn't boring or repetitive. Otherwise, I'd go miles out of my way to avoid it. If it wasn't easy, I might not even try because who knows? I might fail or look silly.

Comfort is sneaky. What seems to be a way to avoid something unpleasant eventually turns into avoiding any challenges. "Why try [insert minor challenge here]? It looks hard." "What if I fail? "Hey, it's too [hot, cold, rainy, windy, sunny, late, early] to work out right now."

Comfort addiction is habitual reliance on physical or emotional ease. It makes you seek out situations or environments that minimize stress, challenge, or discomfort. Major symptoms are the avoidance of difficult tasks and looking for immediate gratification. It's just as dangerous in the long run as substance abuse or gambling.

We don't call out comfort as a real problem. Here's why:

- Cultural acceptance — People see comfort as a positive goal. Marketing campaigns encourage consumers to buy products that make life easier, more convenient, or luxurious.

- No immediate consequences — The bad effects of comfort addiction build slowly over time.

- Invisibility — There are no physical symptoms. Choices that seem harmless (or even beneficial) at first glance, such as binge-watching TV shows, avoiding challenging tasks, or staying in a secure but unfulfilling job.

- Perceived harmlessness — Relaxing on a couch, playing video games, or browsing social media seems normal and harmless. Unlike smoking or drinking, which have well-documented health risks, the dangers of excessive comfort are more abstract and long-term.

- Stagnation — Choosing the path of least resistance makes sure you miss out on opportunities to develop new skills, face challenges, and achieve meaningful goals.

"Illustrations by Dall-E."

Drifting takes you towards the rocks!...

You are the average of your closest friends. Always choosing easy puts you in the company of others who always take the easy way. You'll never be able to build a community of support you can depend on.

Staying too comfortable makes you less resilient, and capable. It ultimately makes you less than you would want yourself to be.

Being too comfortable means staying stagnant. I once bought a guitar and set out to learn how to play it. After a few weeks of starts and stops, I gave the guitar away. It was hard to sound good. I never played consistently enough to toughen my fingertips.

I had studied karate, competed in cross-country, lettered in gymnastics, and started successfully in the Marine Corps. I quit simply because it was harder than I imagined. I did not notice the irony that guitar practice was too tough for the big bad Marine.

Comfort Traps are everywhere.

Marketers make comfort seem as if it's always positive as they push us to buy more to make life easy, luxurious, and safe. The problem is that comfort is fine… until it isn't. You need to eat and drink but who wins when they sell you the supersized BIG MAC COMBO Meal? The extra-large sugar water concoction, all-you-can-eat buffet?

The hidden danger of comfort in the always-on, easily available, and simple-to-consume world of video games, fast food, and never-ending social media scrolling is that **you are wired by default to go past the point where it's healthy for you.**

Leaning in on Discomfort

Most jobs are filled by people who self-select. People who picked a direction, built up credentials, and consistently delivered results filled the slots for all the desirable jobs. Everyone else went where random chance took them.

Learning that lesson allowed me to stop letting my life drift and start paddling in the general direction I wanted to go.

"It's supposed to be hard. If it wasn't hard, everyone would do it. The hard is what makes it great." — Tom Hanks as Jimmy Dugan in "A League of Their Own"

People rate public speaking as one of their number one fears. I guarantee that after you've delivered a hundred speeches, presentations, or pitches, you'll no longer fear it. After a few dozen board-level presentations, I got over it and became the go-to for "interfacing with the board". **Lean in.**

I feared starting my own business and avoided it for years. I got comfortable with the security of my regular paycheck. That comfort kept me from trying until I finally gathered the courage to launch my publishing business. That business eventually failed, but it taught me the importance of embracing discomfort. It also taught me important lessons about running a business. That early failure led to my later success in managing global IT projects for clients like IBM, AT&T, Wells Fargo, and Siemens. **Lean in.**

I've learned a little since I was 12. When I turned 60, I picked up the guitar again, but this time I've been at it for twelve years and I'm a pretty good musician, if only at home. At age 65 I published my first book. At 70, I earned my black belt in Aikido. **I'm leaning in.**

Practical Steps to Lean into Discomfort

- Challenge yourself

- Try new things, even when it's not comfortable.

- Seek out things that are hard to do and learn them.

- Never stop learning.

- Get past the initial discomfort for anything new. If you do something for sixty days and are diligent at it but still don't enjoy it, then stop. You've gotten past the complete beginner stage and it's just not for you.

- Build yourself a community that encourages you to persist through initial discomfort.

It's important to rest, relax, and be comfortable at times. Now is a good example. I believe I'll take a nap. I've finished the work for this post. That was my challenge to myself for today.

Use comfort as a reward, not as a goal.

I wrote this post with assistance from ChatGPT 4o and Claude. Both AIs were asked to criticize my drafts and I used their best suggestions to improve it. Please comment on your thoughts about the results.

That's My Perspective...

08-19-24
Building a Prompt to Improve Your Thinking:

An AI How-To for Dummies

Introduction:

Hello, fellow explorers of the strange things that grab my attention. Here's a straightforward guide on how I built a prompt for my personal Artificial Intelligence assistant (I call him Al, short for Alfred — as in Batman's friend, confidante, and butler). Have you ever wondered how to coax the best out of an AI? Well, buckle up, because today we're diving into the fascinating world of prompt engineering.

Don't worry; we'll keep it simple because **complicated is the fastest way to go wrong**. Think of this as a good starter template.

Step 1: Define Your Core Idea

First things first, you need to have an idea. What do you need the AI to do? State the idea as clearly as possible.

It doesn't need to be grandiose; it could be anything you want to explore. Let's say you want to explore the belief that wearing a hijab is bad for a person. Now, if you're thinking, "Why would I want to question that? It's obvious," stick around – it's all part of teaching the AI to think critically.

Example Belief: "Wearing a hijab is bad for a person."

Step 2: Expand on the Idea

Next, we expand on this belief. We'll explain why some might think it's true. This is where you put on your Sherlock Holmes hat and start digging.

Expansion: Some argue it restricts freedom and imposes gender roles. Sounds like a big deal, right? But hold on, we're just getting started.

Step 3: Break It Down

Now, let's break this idea into its key components. Imagine you're disassembling a Lego model – what are the blocks that make up this belief?

Components: Freedom restriction, gender roles, social pressure. Yep, it's that simple.

Step 4: Seek Counterexamples

Here's where things get interesting. We start looking for situations where this belief might not hold true. Think of it like playing devil's advocate but with a friendly AI sidekick.

Counterexamples:

- **Sun Protection:** Prevents sunburn in sunny climates. (Who doesn't love a good tan, but without the burn?)

- **Social Safety:** Reduces danger in certain cultural contexts. (When in Rome... or, you know, any other place where it's the norm.)

- **Personal Choice:** Empowering for some women as an expression of faith. (Freedom to choose is also a kind of freedom.)

Step 5: Analyze Each Component

Here's where you get to channel your inner scientist. Examine each component individually and see how the counterexamples affect them.

Analysis:

- **Freedom Restriction:** What if it's a personal choice?
- **Gender Roles:** Could it also symbolize strength and identity?
- **Social Pressure:** Is not wearing it riskier in some places?

Step 6: Update Your Understanding

Finally, we synthesize all our findings and update the belief. Think of it as giving your old belief a software update.

Example Belief: "Wearing a hijab is bad for a person."

Updated Belief: "Wearing a hijab has varied implications depending on context and individual choice. It can be both empowering and restrictive."

In other words, **"It Depends," which in this case is a better and more nuanced answer.** You might modify your own set of values to take the social context into account.

The Result: A Standalone Prompt for Naive Users

And voilà! We have a prompt that can guide anyone through the process of deconstructing beliefs.

Everything Below Is Part Of The Prompt Example:

Instructions for Users:

- **Copy and Paste:** Cut and paste the entire prompt into the chat window of a ChatGPT 4.0 or better AI for best results.

- **Important Note:** AI can make mistakes. Always check the results for accuracy and relevance.

Deconstructing Beliefs: A Step-by-Step Guide

Introduction:

- This prompt helps you critically examine and update beliefs by exploring different perspectives and counterexamples.

Steps to Follow:

1. Define the Core Idea:

   - Clearly state the belief or idea you want to examine.

   - Example: "Wearing a hijab is bad for a person."

2. Expand on the Idea:

   - Explain why this belief is considered true.

   - Provide supporting arguments and evidence.

   - Example: Discuss perceived restrictions and cultural implications.

3. Identify Components:

   - Break down the belief into key components.

   - Example Components: Freedom restriction, gender roles, social pressure.

4. Seek Counterexamples:

   - Look for situations where the belief may not hold true.

   - Provide counterexamples.

   - Example: Sun protection, social safety, personal choice.

5. Analyze Each Component:

   - Examine the validity of each component individually.

- Assess the impact of counterexamples on the overall belief.
- Example: Explore how wearing a hijab can be empowering or restrictive.

6. Update Understanding:

- Synthesize your findings to update the belief or statement.
- Formulate a revised understanding based on your analysis.
- Example: "Wearing a hijab has varied implications depending on context and individual choice. It can be both empowering and restrictive."
- Conclusion: By following these steps, you can develop a more nuanced and updated understanding of beliefs. Use this process to critically examine various ideas and improve your thinking.

Conclusion:

And there you have it! A simple and hopefully enlightening guide to building a new prompt to challenge any sort of idea. So next time you find yourself in a philosophical debate or just curious about a belief, you know how to deconstruct it like a pro.

Special Thanks: Big thanks to Al, my ChatGPT AI assistant, for helping me make this post clearer and more engaging. Extra special thanks to Donna, my editor, for keeping the whole thing real.

That's My Perspective...

08-26-24
Soft Skills: Life Beyond the Classroom

If I knew then what I know now...

Introduction:

This post explores the crucial "soft skills" that go beyond the classroom and technical training, skills that are essential for navigating both professional and personal life.

Becoming a new project manager felt like building a race car in the middle of a race. Trying to tune the engine as the wheels fell off.

"Illustrations by Dall-E."

Building the car in the middle of the race

You take your first sip of coffee as you log in to see what fires have moved to the front burner. Fuzzy and trying hard to get up to speed as the first email hits you—it's one of one hundred twenty-six this fine morning. Thirty-one are marked urgent. The deadline for your next major milestone is in two days, and your voicemail is full. It's Wednesday.

The Problem:

School did not prepare me for the real-world challenges of corporate project management. Swimming upstream left no room for error or downtime. "I'm constantly putting out fires and barely keeping up. When will they realize how bad I am and fire me?" The pressure was immense, and my technical skills alone weren't going to be enough.

My communication, decision-making under pressure, and negotiation skills were crap. These weren't just useful— they were essential. The challenge wasn't just about "getting the job done" but also about mastering the "soft skills" I needed to survive.

The Learning Curve:

I knew some areas giving me pain, so I started to fix them. I began to see progress. Each new skill felt like adding a sharper tool to my toolbox. Before long, I realized that I wasn't drowning anymore.

Reading David Allen's *Getting Things Done* gave me the breathing space to start. "**Your mind is for having ideas, not holding them.**" This principle helped me organize my thoughts and tasks, allowing me to focus on action rather than getting bogged down by clutter. The atmosphere at work began to shift. What once felt like an avalanche was still an avalanche, but I was starting to learn how to surf on top. The chaos was still there, but I was getting better at riding it out.

Being willing to learn was the key. Organization, time management, business, negotiation, communication, teamwork, public speaking, and the most important meta-skill of all—learning how to learn and teach myself new skills—became my new targets. The more I practiced, the better I got at handling difficult assignments. That built my confidence based on real accomplishments.

The Broader Impact:

As I gained skills, I changed my approach to work and home life. Impossible challenges became manageable, and people began to trust my judgment. My confidence grew. But as my boss would say, "No good deed goes unpunished."

As soon as I finished a large, complex project, I'd be assigned something much larger and more complex. He pointed out, "We're only as good as our most recent project," and assigned me several failed projects to rescue. As I became more effective, I went from small telecom installations to global projects with teams in India, China, the EU, and on both coasts.

Soft skills affect every part of our lives, from planning a birthday party to tactful and empathetic personal relationships. The ability to communicate well, make good decisions, and negotiate effectively are the essential everyday skills I never practiced in school.

"If you are serious about changing your life, you'll find a way. If you're not, you'll find an excuse."

— *Learning How to Learn* by Barbara Oakley and Terrence Sejnowski

Remember this:

The way to thrive is to become a lifelong learner. Going from struggling with basic survival in the corporate world to managing complex, global projects drove home the one thing you should take away from this post.

The ability to adapt, learn, and improve isn't just useful— it's essential. Soft skills aren't just for work, they're life skills. The so-called "soft skills" are anything but soft. They're the backbone of effective leadership and personal growth.

What skills do you need to build? Find your weaknesses (and strengths) so you can shore up the weak spots and double down on the strengths. Personal and professional growth is a lifelong commitment, and there's always room to improve.

Special Thanks:

Special Thanks: Big thanks to Al, my ChatGPT AI assistant, for helping me make this post clearer and more engaging. Illustrations also courtesy of Al.

Post-Script:

Extra Insights:

Getting Things Done by David Allen: "Your mind is for having ideas, not holding them." *Learning How to Learn* by Barbara Oakley and Terrence Sejnowski: "If you are serious about changing your life, you'll find a way. If you're not, you'll find an excuse.

That's My Perspective...

09-02-24
Filling the Gaps:

The Small Fixes That Make a Big Difference

"**This house is miserable!** The kitchen feels like a furnace, the bedroom and bathroom are freezing, my skin is cracking from the dryness, and drafts are creeping in from every corner." That was my wife Donna's take on the house we bought in May. Now that winter had arrived, our new home was expensive to heat, freezing in some rooms, and a sauna in others.

We considered new windows, doors, insulation, and siding, but the costs were prohibitive. So, we held off on any major changes until we could save enough to afford them. Meanwhile, we made do with extra blankets, thick socks, and a space heater at night.

At the same time, I was embarking on a new venture—starting an energy auditing company. I got certified as an energy auditor and home inspector, ready to help others improve their homes. But with all that going on, I never got around to auditing our house.

That winter, Donna pointed out the obvious, so I set aside time for an energy audit and inspection, expecting the worst—needing insulation, siding, new windows, and doors.

Instead, I discovered that small, strategic fixes had a greater impact on our home's comfort and efficiency than I ever imagined.

By the following summer, our air-conditioning bill was a third lower. When winter returned, the whole house was free of hotspots and drafts. Every room in the house was comfortable, and we decided that replacing the windows or siding wasn't necessary (though we did add insulation in the attic). Our heating bill was halved.

"Illustrations by Dall-E."

The Leaky Bucket:

What magic made the drastic difference? It was the simplest and cheapest fix—but first, let me explain.

My guitar teacher and friend, Roger Pitts, used the metaphor of a "Leaky Bucket" during one of our lessons. He was talking about guitar practice, but it applies to much more. Photo by Ron Lach

Everyone has leaks. Small things we do, even when we know we shouldn't, that work against us. We forget things we learned, we pick up bad habits, we get lazy, and even skills we've mastered will degrade over time.

There are ways to deal with this. You can add more practice, like pouring water into the bucket faster than it can leak out. But realistically, who has four hours a day to practice guitar?

Or, you can patch the leaks so the water you add doesn't spill. In guitar practice, or any learning process, a method called "Mastery Learning" helps patch those memory leaks. It's the gold standard for learning. Just like a child who struggles with basic math will find algebra difficult, not gaining mastery before moving on leaves you with a leaky bucket.

Patching the Leaks:

We hear from every "self-improvement" or "healthy living" guru that there are some basics we must do to thrive: e.g. diet, exercise, sleep, and community. But here's the thing—before you try to optimize part of your life, stop letting leaks undo all your hard work.

Have you heard of the **latest and greatest diet?** Whether the new diet is keto, gluten-free, organic, vegan, or Mediterranean, get the basics right first. Eat a balanced diet with lots of non-starchy vegetables, drink plenty of water, eat slowly, and stop when you're full. *Patch the leaks before you optimize.*

The latest exercise fad? Great. But first, make sure you're getting the basics: move every day, stretch, lift some heavy stuff a couple of times a week. Quit smoking and vaping. *Patch the leaks before you dive into the latest craze.*

Sleep advice? It's all about consistency: get at least eight hours, wake up at the same time every day, and sleep in a cool, dark room. Fancy sleep aids and routines can help, but they won't fix a leaky bucket. *Patch the leaks first.*

Community? Take the time to reach out to friends, family, and community. Social connections are the glue that holds life together. Social media can't replace real human interaction. *Patch the leaks first.*

Hard decisions? Take care to step back and make sure of the basics first. Are you solving the right problem? Have you checked the risks and eliminated options that could end in disaster? Have you considered what happens next? *Fix the leaks.*

The Simple Fixes:

So, what were the "magical" fixes that transformed our home? The energy audit revealed a disconnected HVAC duct in the ceiling. Fixing it took about twenty minutes and cost nothing. It allowed the system to properly heat and cool the bedroom and bath, reducing the excessive heat in the kitchen and stopping the attic from getting overheated.

"Illustrations by Dall-E."

I then spent a weekend caulking and weatherstripping the house. I sealed and insulated the rim joist, which runs around the perimeter of the floor system like a belt. It's usually leaky in older homes, and ours was no exception. The weekend's work and a couple of hundred dollars made the difference.

Before your next big thing, focus on filling the gaps. Getting the basics right makes every effort more effective, whether in your home, your health, or your life.

Special Thanks:

Thanks to Al, my ChatGPT AI assistant, for helping me make this post clearer and more engaging.

That's My Perspective...

09-09-24
Brainstorming Busts

Avoid failure modes

Today's post will help you get the most out of your brainstorming sessions.

Brainstorming can go completely wrong. It's meant to bring out the best ideas from a team and explore a wide range of possibilities. Done well, it builds the team while exploring lots of approaches. On the other hand…

"Illustrations by Dall-E."

You're in the office, meeting the team for the company's next big project. Most of them are offsite, dialed into a conference bridge. You only know a few people. This project is make-or-break for your division. Nervous chatter fills the room as people introduce themselves and find their chairs. Your last project didn't go well, so you're feeling jittery.

The morning sun is blinding the people across from you, so you stand up and pull the blinds to murmurs of appreciation. It feels like that might be the most appreciation you'll get all day.

Your mission today is to get this team up to speed on a project that could be overdue before the budget is even approved. You've received vague directions from the executive who 'owns' this project—a guy known for his 'hands-off' approach.

In other words:

1. You don't know exactly what you're supposed to deliver—just a vague idea of the direction, but nothing final about the end state.

2. You have a rough budget estimate but can't spend anything without executive approval.

3. You're leading a new team you've never met.

4. All the team members report to different departments and managers, and, for the duration of this project, also to you. Each of them has other responsibilities aside from this project.

5. The deadline is approaching fast.

Welcome to project management.

The deep end.

I started a project like that early in my career. I thought, 'This kickoff meeting needs an icebreaker to get people acquainted. After a quick introduction, we'll dive into a brainstorming session.'

Rule #1: You must establish psychological safety first.

Without realizing it, I went directly against the most important rule for brainstorming.

The team members came from a cross-section of departments across the company. A few were assigned full-time to the project, but most had tasks scattered across different areas. Some departments were rivals, used to working in silos without worrying about cross-functional collaboration. Since we were at headquarters, the people in the room were high-ranking managers or executives, while those who dialed in were technicians, engineers, and local representatives from our scattered offices.

"Now that I've outlined the situation, let's see if this group can brainstorm ideas about how to proceed," I thought optimistically.

Crickets…

"Alright, team, we've got a lot of talent here. Surely someone has some ideas. Nothing is too far-fetched or silly."

Bigger crickets…

By this point, I was worried my nervousness was starting to show, so I tried to steer the conversation. "Maybe we can start by suggesting ways to break the project into manageable chunks?" I offered.

A voice from the phone asked, "Isn't that what the project plan is for?" From that point on, the meeting became an awkward exercise in prying information from people who didn't want to share.

This meeting was a "learning opportunity" for me. My mistake? I had set up the kickoff meeting in a way that actively discouraged sharing.

Several things I did wrong:

1. While "the boss" was in the room, most people waited to hear their opinions. I had several executives in the mix.

2. The entire team had about 30 people—far too many strangers to feel comfortable throwing out goofy ideas.

3. The team had a history together, but it was a competitive one.

4. I didn't have a well-defined problem to solve, just a general question.

5. I was asking people to speak out in a group setting without any assurance that it was safe to bring up the things we needed to discuss.

6. I hadn't thought about how much credence to give each speaker.

   The next day, my boss pulled me aside to give me some "feedback" about the brainstorming session. I braced for the worst…

   He said, "Aside from the brainstorming session at the start of the kickoff, I heard things went well."

   I braced myself for the typical management "compliment sandwich"—start with something nice, then keelhaul you, and finish with "keep up the good work."

   Instead, I got: "I understand the brainstorming didn't go well. What would you change next time?" He followed it with an expectant look—no recriminations, no scowling, just a straightforward question.

   I had been stewing over it, focusing on the lack of participation from the team. I hadn't considered what I could do to change the results.

   Then he surprised me. "Don't try to answer that right away," he said. "Think about it and let me know what you come up with on Monday." With that, he let me get back to work.

Monday's assignment looked like this:

1. Don't use brainstorming as an ice-breaker. People won't open up to others they don't know.

2. Keep the bosses out of the room if you want everyone to speak candidly.

3. Exclude the workers if you need to discuss strategic planning with the executives.

4. Keep the group small—it's easier to talk in front of a few people.

5. Have a well-defined problem to solve or start by defining the problem.

When I met with him on Monday, he reviewed my responses and added these thoughts:

1. Instead of having everyone speak their answers aloud, give them 3 to 5 minutes to write as many responses as possible. Then, collect and share them anonymously on a whiteboard. This way, each idea gets a hearing without anyone feeling exposed.

2. Ensure the right mix of people in the session. It's unproductive to have executives weigh in on deeply technical problems or involve someone with no "skin in the game."

3. When consulting experts, don't start by asking them to solve the problem. Ask them how they think about it. Gather input from two or more experts and have them combine their methods. Use that as a framework for solving the problem.

He then sent me back into the fray. No drama.

I had just gotten a great lesson in coaching. Give honest feedback, challenge people to think about it, and then let them learn and grow.

Since then, my brainstorming sessions have improved. I've even added these gems:

1. Start any brainstorming session with a "safety" introduction.

2. Tell people the purpose of the meeting is to stir up ideas, not to evaluate them.

3. Encourage them to produce the goofiest ideas possible, with extra points for making the team laugh.

4. Treat it like an improv exercise, where any starting point should be followed by "Yes, AND."

5. When you evaluate the ideas produced, do it in a separate session.

People need to feel safe expressing their ideas. When they feel safe and valued, they will share.

For me, the lesson was in two parts. My boss demonstrated how to give constructive feedback respectfully, treating me like a peer and offering tips rather than the browbeating I expected. He also made me examine how I could change my approach to get better results. By making me responsible for my improvement, he gave me agency and let me choose how to "own" my career.

Both of those lessons have served me well.

That's My Perspective…

09-16-24
You are the hero of this story...

"Have you ever felt fear, uncertainty, doubt, or shame weigh
   you down? Life's villains wear many masks — anxiety,
   social awkwardness, poverty, and even self-doubt. If
   you've struggled with these, you're not alone. And if
   you've ever felt like an impostor, skating by, wondering
   "is today the day they find me out?" ... well, maybe you're
   right.

Or maybe it's just for today. Because guess what? There's a
   cure.

No quick fixes or miracle cures here—no magical thinking or
   new religion. But there's a real approach that works, as
   long as you're willing to learn and put it into practice. If
   you're ready to dive into the unknown, let's go.

"Illustrations by Dall-E."

Dive into the unknown

'When will I ever use this?' Remember sitting in school, staring out the classroom window and daydreaming while you were supposed to be slogging through algebra or memorizing useless dates. It felt pointless, didn't it? As if life's real lessons were hiding somewhere else, waiting to ambush you when you least expected it.

"I wish I knew back then what I know now" — (Said by everyone, ever.)

This is your chance to learn NOW what you will need THEN for the rest of your life.

Enough teasing. Here's the truth: The decisions you make every single day—whether small or life-altering—determine everything. And the way we often feel lost or overwhelmed by choice? That can change. Imagine a framework to guide your decisions, whether it's everyday choices or high-stakes moments.

Better Decisions Make a Better Life.

This series is your roadmap to mastering decision-making. Imagine a course that teaches you how to make the decisions that truly shape your life—choices about your career, relationships, and even your personal growth. That's what this series offers. It's the class you wish you'd taken years ago.

As we go, we'll break down powerful strategies—everything from handling everyday low-stakes choices with ease to navigating high-pressure, decisions with clarity and confidence. By the end of this journey, you'll have the skills to make confident, informed choices in any situation, no matter how challenging.

1. Purpose and Values: Establishing Your Foundation

Description: Effective decision-making begins with a clear understanding of your personal purpose and core values. Knowing what you stand for and what you aim to achieve provides a compass that guides all your choices.

- **Aligning Decisions with Core Beliefs:** Reflect on your fundamental principles and ensure your decisions are consistent with them.
- **Defining Your Mission:** Articulate your overarching goals to provide direction and meaning.

Recommended Reading:

"Discovering Your Authentic Core Values," *Psychology Today*. Read more here

Frankl, Viktor E. *Man's Search for Meaning*. Beacon Press, 2006.

2. Self-Awareness: Understanding Yourself

Description: Self-awareness is crucial for recognizing how your emotions, biases, and thought patterns influence your decisions. Developing emotional intelligence allows you to manage your reactions, especially under stress.

- **Emotional Intelligence:** Learn to identify and regulate your emotions.
- **Cognitive Biases:** Recognize common biases that can cloud judgment.

Recommended Reading:

"Emotional Intelligence: Why It Can Matter More Than IQ," Daniel Goleman. Read more here

Kahneman, Daniel. *Thinking, Fast and Slow*. Farrar, Straus, and Giroux, 2011.

3. Preparation: Laying the Groundwork

Description: Good decisions are built on solid preparation. This involves gathering relevant information, understanding your goals, and aligning your actions with your values. Being mentally and emotionally ready to decide is key.

- **Information Gathering:** Collect data from reliable sources to inform your decisions.
- **Goal Setting:** Clarify what you want to achieve with each decision.

Recommended Reading:

"Simple Steps for Better Decision-Making," *Psychology Today*. Read more here

Heath, Chip & Dan. Decisive: How to Make Better Choices in Life and Work. Crown Business, 2013.

4. Triage: Prioritizing Decisions

Description: Not all decisions carry the same weight. Triage helps you prioritize, focusing your time and energy on what matters most. Learning to differentiate between urgent, important, and trivial decisions is a game-changer.

- **Urgent vs. Important:** Identify tasks that require immediate attention versus those that are important but not time-sensitive.
- **Resource Allocation:** Allocate your time and mental energy appropriately.

Recommended Reading:

"Decision-Making Under Stress," *American Psychological Association*. Read more here

Bazerman, Max H. Judgment in Managerial Decision-Making. Wiley, 2012.

5. Decision-Making Frameworks: Choosing Wisely

Description: Employ structured approaches to make informed decisions. Utilize rational models, intuitive insights, or a hybrid to evaluate your options thoroughly.

- **Analytical Tools:** Use decision trees, cost-benefit analysis, and probability assessments.
- **Intuitive Thinking:** Leverage your experience and instincts when appropriate.

Recommended Reading:

"How to Create a Decision Journal," Farnham Street Blog. Read more here

Hammond, John S., et al. "The Hidden Traps in Decision Making," *Harvard Business Review*, 1998.

6. Stress Management: Maintaining Clarity Under Pressure

Description: High-stress situations can impair judgment. Learning techniques to manage stress ensures you maintain mental clarity when making critical decisions.

- **Mindfulness and Relaxation:** Practice techniques like deep breathing and meditation to stay calm.
- **Resilience Building:** Develop coping strategies to handle pressure effectively.

Recommended Reading:

"Stress Management Techniques," *Mayo Clinic*. Read more here

Selye, Hans. *The Stress of Life*. McGraw-Hill, 1976.

7. Critical Thinking Skills: Enhancing Your Analysis

Description: Critical thinking enables you to analyze situations objectively, consider multiple perspectives, and avoid common pitfalls in reasoning.

- **Problem-Solving Techniques:** Use methods like root cause analysis to address underlying issues.

- **Avoiding Logical Fallacies:** Recognize and steer clear of flawed arguments.

Recommended Reading:

"Critical Thinking Skills," *MindTools*. Read more here

Paul, Richard, and Linda Elder. Critical Thinking: Tools for Taking Charge of Your Professional and Personal Life. Pearson, 2013.

8. Ethical Decision-Making: Making Principled Choices

Description: Ethical decision-making ensures your choices align with your personal values. It involves considering the broader impact of your decisions.

- **Ethical Frameworks:** Understand theories and choose your approach.

- **Stakeholder Impact:** Consider how your decisions affect others.

Recommended Reading:

"A Framework for Ethical Decision Making," Markkula Center for Applied Ethics. Read more here

Rest, James R. Moral Development: Advances in Research and Theory. Praeger, 1986.

9. Results and Review: Understanding Outcomes

Description: After making a decision, it's crucial to evaluate the results. Avoid the trap of "resulting"—judging the decision solely by its outcome. Instead, focus on how the decision was made and what can be learned from the process.

- **Process Evaluation:** Assess the quality of your decision-making process.
- **Outcome Analysis:** Understand the results to inform future decisions.

Recommended Reading:

"Resulting and Decision Quality," Farnam Street Blog. Read more here

Russo, J. Edward, and Paul J. H. Schoemaker. *Winning Decisions*. Crown Business, 2002.

10. Reflection and Debriefing: Learning and Adjusting

Description: Reflection helps lock in progress and prevent repeated mistakes. A formal debrief or post-mortem allows you to dissect the decision-making process and improve future strategies.

- **Self-Assessment:** Regularly evaluate your decisions and thought processes.
- **Feedback Integration:** Incorporate lessons learned into your decision-making framework.

Recommended Reading:

"How to Reflect on Your Decisions," MindTools. Read more here

Schön, Donald A. The Reflective Practitioner: How Professionals Think in Action. Basic Books, 1983.

11. Communication Skills: Articulating and Influencing

Description: Effective communication is essential for decision-making, especially when others are involved. It ensures that your ideas are understood and that you can influence outcomes positively.

- **Active Listening:** Fully understand others' perspectives and concerns.
- **Clear Expression:** Communicate your decisions and reasoning transparently.

Recommended Reading:

"Effective Communication," *SkillsYouNeed*. Read more here

Cialdini, Robert B. *Influence: The Psychology of Persuasion*. Harper Business, 2006.

12. Continuous Improvement: Enhancing Your Decision-Making Framework

Description: The final step is gathering feedback to improve your decision-making process. Like a ratchet mechanism locks in progress, continuous improvement helps secure gains and propels you forward.

- **Feedback Loops:** Create mechanisms for receiving and acting on feedback.
- **Adaptability:** Stay flexible and open to change as new information emerges.

Recommended Reading:

"The Importance of Feedback in Decision-Making," *Harvard Business Review*. Read more here

Popper, Karl. The Logic of Scientific Discovery. Routledge, 2002.

Conclusion

We've now outlined the key concepts that will guide your decision-making journey. These ideas form the backbone of making choices effectively—whether under everyday circumstances or in high-pressure situations. By starting with a solid foundation of purpose and values and building through self-awareness, preparation, and strategic action, you'll be better equipped to make decisions confidently and competently.

In the coming posts, we'll dive into each of these topics in more detail, interspersing them with other subjects that matter to you. We'll explore practical tools, real-world examples, and strategies to enhance your decision-making skills.

For now, reflect on your own approach:

- How do you prepare for decisions?
- Do you understand your core values and how they influence your choices?
- Are you aware of your cognitive biases?

Consider starting a decision journal to track your progress—it's an invaluable tool as we continue to explore these concepts together.

Special Thanks

Big thanks to Al, my ChatGPT AI assistant, for helping me make this post clearer and more engaging.

Post-Script

I am one of those people who gets obsessive about something and really dives in deep. I've been obsessed with good decision-making for decades. However, that doesn't mean I won't also be covering other interesting topics. I plan to sprinkle in whatever has really grabbed my attention, so we don't all go stale on decision-making topics.

That's My Perspective...

09-23-24
Purpose and Values:
Establishing Your Foundation

First Principles

Don't go off half-cocked, half-baked, or half-assed.

- Tony Pray

"Illustrations by Dall-E."

Effective decision-making has to start by going in the right direction. If it doesn't line up with your purposes or if it contradicts what you value most, it can't be right for you.

Establish your "Why"

Every time I tried to think about my "purpose" in life, my mind would veer off into distractions. It took me a very long time to finally cut through the B.S. I was selling myself to get to the core of what makes me tick. Your results will be different.

I believe it's worth the effort. If you don't know the goal, your efforts are likely to be in the wrong direction. If you have a purpose in mind, you will at least be going in the right general direction.

"He who has a *why* to live can bear almost any *how*," — Victor Frankl

Determine your "values".

Next, you need to **understand what you value**. Is it 'family", "loyalty", "friendship", "power", "peace", "honesty", "popularity", "money" or any combination of attributes? **What are the best things you aspire to be?** Your values are your roadmap to the kind of person you want to be while you go after your goals. Maybe your purpose in life is to gain power. If you value subterfuge, you could be a Machiavelli. If you value cooperation, you might be an Andy Reid (KC Chiefs head coach). Both got power, in ways guided by the things they value. **Your values determine *how* you make the journey.**

Knowing where you're going and how you'll get there is the foundation of great decision-making.

The following recommendation is not a homework assignment, just a great source if you want to dig deeper.

Frankl, Viktor E. *Man's Search for Meaning*. Beacon Press, 2006.

That's My Perspective...

09-30-24
Self-Awareness:

You aren't exactly what you think...

Self-Awareness: Hidden Biases and Emotional Intelligence

Your mind is playing tricks on you. It's shaping your decisions without you even noticing!

Self-awareness isn't just about knowing your strengths and weaknesses—it's about recognizing when to slow down and give your conscious mind time to triage. Sometimes, trusting your gut is fine, but there are moments when you need to step back and decide if a thoughtful, deliberate choice is necessary. Two essential, but often overlooked, areas in this journey are hidden biases, especially confirmation bias, and emotional intelligence. Let's explore these concepts to boost your self-awareness effectively.

1. The Subtle Influence of Confirmation Bias

Confirmation bias is the tendency to favor information that aligns with your existing beliefs while ignoring contradictory evidence. It quietly influences your decisions, often without you realizing it.

Imagine you're debating with a friend about the best way to stay healthy. You've always believed that a low-carb diet is the way to go, and you've read plenty of articles that back it up. During the debate, your friend presents several studies showing that a balanced diet, including carbs, leads to long-term health benefits. Instead of weighing their arguments fairly, you zero in on one line in their research that mentions a potential short-term benefit of reducing carbs, completely ignoring the overwhelming evidence for a balanced approach. You leave the conversation even more convinced that your low-carb diet is right—your mind filters out everything that doesn't fit your belief.

Now think about trying to discuss something sensitive, like gun control, abortion, or politics. **Two people can hear the same set of facts and still walk away with completely different opinions**—each latching onto the parts that fit what they already believe and brushing off the rest.

- **Impact on Self-Awareness**: Bias can prevent honest self-reflection. You may overlook mistakes or credit your success only to yourself, ignoring external factors.

- **Overcoming Confirmation Bias**: Counteract bias by seeking opposing viewpoints and engaging with challengers. This broadens your understanding and encourages critical thinking.

2. Deepening Emotional Intelligence

Emotional intelligence is all about recognizing, understanding, and managing both your own emotions and the emotions of others. It's like the skill of reading the room—only, the "room" is your inner self.

- **Emotional Granularity**: This is the ability to accurately label your emotions. Instead of just feeling "bad" or "good," you can pinpoint specific feelings like frustration, disappointment, or excitement. This clarity helps you manage emotions better.

- **Emotional Agility**: It's about **accepting your emotions without letting them control your actions**. Know when to pause so your conscious mind (the "monkey") can better manage the powerful emotions driven by your subconscious (the "elephant").

Understanding and developing emotional intelligence allows you to see beyond surface reactions, addressing deeper issues and avoiding knee-jerk responses. The more skilled you are at this, the more self-aware you become.

3. The Monkey and the Elephant: Conscious Control vs. Subconscious Power

Think of your mind like two minds in one: the conscious, calculating monkey and the instinctive, powerful elephant.

- **The Monkey**: The monkey represents your conscious mind—the part of you that makes thoughtful, deliberate decisions. It analyzes situations, weighs options, and tries to steer you in the right direction.

"Illustrations by Dall-E."

- **The Elephant**: The elephant is your subconscious. It's far stronger and driven by instinct, emotions, and automatic reactions. When the elephant is calm, it quietly follows the monkey's lead. **But when it gets upset, the elephant can easily overpower the monkey,** charging forward based on raw emotion.

Self-awareness means learning to recognize when the elephant is about to take over and knowing how to calm it. This is where emotional intelligence comes in—it helps the monkey manage the elephant, ensuring that your decisions are thoughtful and not just reactions to intense emotions

WHOA!

4. Putting It All Together:

Understanding how confirmation bias and emotional intelligence influence the monkey and elephant dynamic is key to improving your self-awareness.

- **Biases and the Elephant**: Confirmation bias fuels the elephant. The subconscious craves comfort and certainty, so it filters out information that challenges your beliefs. When the elephant senses something it doesn't like, it charges ahead, ignoring the monkey's attempts to think things through.

- **Emotional Intelligence and the Monkey**: Emotional intelligence empowers the monkey. By recognizing and managing emotions, the monkey can calm the elephant, giving it time to assess the situation and make more rational decisions.

When biases take control, your subconscious elephant reacts on impulse, but emotional intelligence lets your conscious monkey take back the reins. Mastering this balance leads to better self-awareness and decision-making.

5. Practical Tips to Boost Self-Awareness:

Here are some simple yet powerful strategies to help you uncover hidden biases and enhance emotional intelligence:

- **Mindful Reflection and Journaling**: Set aside time each day to reflect on your thoughts and emotions. Write down what triggered your emotions and ask yourself whether any biases influenced your reactions. This helps you spot patterns and improve emotional management.

- **Get Feedback from Different Perspectives**: Talk to people with diverse viewpoints and ask for feedback. Hearing how others see the situation can help you reduce confirmation bias and uncover blind spots.

- **Continuous Learning**: Read up on cognitive biases and emotional intelligence. Books, workshops, and online resources can offer valuable techniques to further develop your self-awareness.

Improving self-awareness isn't just about knowing your strengths and weaknesses—it's about recognizing when your subconscious (the elephant) is steering your decisions and when to let your conscious mind (the monkey) take charge. Understanding and addressing hidden biases like confirmation bias and developing emotional intelligence gives you a more accurate view of yourself.

Further Resources

- Books:
- Thinking, Fast and Slow by Daniel Kahneman
- Emotional Agility by Susan David
- Emotional Intelligence: Why It Can Matter More Than IQ. By Daniel Goleman
- Tools & Apps:
- *Headspace*: For guided mindfulness and meditation practices.
- *Daylio*: A journaling app to track moods and activities.

That's My Perspective...

10-07-24
Laying The Groundwork:

How to Prepare for Great Decisions

Be Prepared

Great decisions don't happen by accident. They're built on smart prep. Whether you're deciding what to eat or making a career move, it starts with the right foundation. Let's break down the essentials: asking the right questions, gathering info, aligning with your values, and being mentally fit to make the call.

Ask the Right Questions

Before any decision, clarity is key. The best way to get clear is by asking the right questions. Start with: **What am I trying to achieve?** Once you define the goal, the path becomes easier to navigate.

- **Is the goal clear?** If your objective is vague, you're setting yourself up for second-guessing. Nail down the specifics before moving forward.

- **What does success look like?** Picture the ideal outcome. A clear vision of success helps you measure progress and stay on track.

- **What's driving this decision? Why are you doing this?** Understanding the "why" behind the decision helps identify the problem you're solving. Without clarity, you risk fixing the wrong issue. Focus on the core problem to avoid distractions.

By asking focused, purposeful questions, you strip away distractions and give yourself the best chance of reaching a confident decision.

Gather Information

Great decisions rely on solid information. But it's not just about collecting data—it's about **asking the right questions** to get what you need. Start by identifying the key facts required to move forward.

- **What do you need to know?** Focus on long-term results and the information that directly impacts the decision. Example: If you're buying a car, prioritize reliability and cost of ownership over short-term features like color.

- **Where can you find it?** Instead of asking experts to solve the problem, ask how they would approach it and what information they would seek. This helps you gather key insights without inheriting their biases.

- **Are you asking the right questions?** Make sure your inquiries are specific. Example: Instead of asking, "Is this product good?" ask, "How will this product improve my workflow?"

Efficient information gathering is about targeting what matters most, filtering out the noise, and ensuring you have what you need to make an informed call. **Stop when you have enough to make a decision confidently—don't wait for perfect information that may never come.**

When to Stop Information Gathering

For big decisions, the temptation to keep searching for more information can be overwhelming, but it's essential to know when to stop. A few rules of thumb:

- Stop when further delay would limit your options.

- Stop when new information would need to be extremely negative to change your mind.

- Weigh the probabilities and trigger the decision once you've reached a certain threshold of certainty.

Avoid "analysis paralysis" and keep moving.

Aligning with Your Purpose and Values

The best decisions align with who you are and what truly matters to you. If a choice doesn't sit right ethically, it's not an option—period.

- **Does it feel right?** If it even slightly feels off or conflicts with your core beliefs, walk away. If it smells fishy, it's not worth the risk.

- **Will this move you closer to your long-term goals?** Every decision should push you toward your bigger purpose. If it derails you, reconsider.

"Illustrations by Dall-E."

Something fishy here!

Aligning decisions with your purpose makes sure you're moving toward meaningful goals, not just chasing quick wins or avoiding discomfort. **Position yourself for the long term.**

Be Fit to Decide

You can have all the right information, but if you're not in the right mental or emotional state, your decision can suffer. Don't make decisions in HASTE. The **H.A.S.T.E.** rule: You're not fit to decide if you're Hungry, Angry, Stressed, Tired, or Excited.

- **Hungry:** If your body is low on fuel, your brain will be too. Eat something before making any big call.

- **Angry:** Anger clouds judgment and narrows your focus. Wait until you've calmed down to ensure you're thinking clearly.

- **Stressed:** Stress leads to impulsive decisions. Recognize when you're overwhelmed and take a break to avoid rushing into something you'll regret.

- **Tired:** Fatigue dulls your ability to weigh options. Sleep on it if necessary, and make sure you're rested before deciding.

- **Excited:** Excitement can be just as dangerous as stress—don't let it sweep you into hasty decisions. Pause, reflect, and make sure your judgment is sound.

By using the H.A.S.T.E. rule, you ensure that your decisions are made with a clear mind and balanced emotions, instead of reacting to your immediate state.

Rules of Thumb for Low-Impact Choices

Not every decision requires deep thinking. For every day, low-impact choices, having simple rules of thumb can save time and mental energy while improving your odds of making good decisions. The key is to **pre-commit to choices that align with your goals**, making follow-through easier and defensible if questioned.

- **Set clear limits.** *Example: "Only one dessert a week."* This removes the need for daily decisions about sweets and keeps you on track with your health goals.

- **Simplify spending.** *Example: "No impulse buys over $50."* This prevents financial stress and forces you to think before making bigger purchases.

- **Streamline routines.** *Example: "Exercise for at least 15 minutes every morning."* By making it a non-negotiable part of your day, you cut out the decision fatigue around when or how much to exercise.

- **Focus on relationships.** *Example: "Call one family member every weekend."* This keeps you connected without needing to plan out every interaction.

These rules of thumb are designed to make your decision-making easier and more reliable, boosting the odds that you'll consistently make choices that align with your broader goals.

Great decisions don't happen by chance—they're the result of preparation, clarity, and alignment with your values. By asking the right questions, gathering essential information, and ensuring you're mentally fit to decide, you lay the groundwork for success. Implement simple rules of thumb for everyday choices and stay committed to decisions that serve your long-term goals.

When you're prepared and equipped with the right mindset, you're in the best position to **make decisions you can confidently stand by.**

That's My Perspective...

10-14-24
Emergency Triage for Decisions

Smarter, Faster Choices...

First: Don't make decisions in H.A.S.T.E. (Hungry, Angry, Sad, Tired, Excited) – pause, check in with yourself, and decide what kind of attention this decision needs."

Triage is about prioritizing wisely, but to do that, you need to be in the right headspace. In bad times, it's easy to make snap decisions. Know when to slow down and assess. Triage helps you sort and prioritize decisions, not just in emergencies, but in everyday life.

"Illustrations by Dall-E."

What is Triage?

Triage is used in medical emergencies to sort patients by urgency. The idea is simple: focus on what needs immediate attention and what can wait. But triage isn't just for life-or-death situations—it's crucial for everyday decision-making. Use it to decide how much effort each decision deserves.

Before You Dive In...

Not every decision is as simple as it looks. A small choice can have big consequences, and a complex one might just need a quick pro-con list. Before you dive in, **pause**. Most decisions are straightforward, but some require more thought. The hardest ones combine emotional weight with complexity. These need the most careful attention—don't rush in.

"Triage is the art of making decisions under pressure; it's not just about sorting the urgent from the important, but about saving lives with every choice."

Types of Decisions: From:

"Do I eat dessert?" to: BIG Life Choices

When triaging a decision, the key factor to consider is **potential impact**. This includes both positive and negative outcomes, and how the decision affects you, the other stakeholders, and even broader society. While **impact** is the primary guide, **urgency** and **complexity** also play important roles.

To visualize this, think of the **Eisenhower Matrix**, which breaks decisions into four categories based on urgency and importance:

- **Low urgency, low importance** (quick decisions with minimal impact).
- **Low urgency, high importance** (long-term decisions that can be planned carefully).

- **High urgency, low importance** (minor decisions that need quick resolution).

- **High urgency, high importance** (critical decisions requiring immediate attention and careful thought).

EISENHOWER MATRIX - EXAMPLE

	Urgent	Not Urgent
Important	**1 DO FIRST** • Repair your broken down car • Pick up your ill child from school • Handle an urgent client request • Cover for a sick colleague • Finish a report due for submission today	**SCHEDULE 2** • Planning and organizing your workload • Strategic planning • Career planning • Attending training courses • Routine maintenance tasks • Regular self-care
Not Important	**DELEGATE** • Urgent factory equipment repair • Requests for help from work colleagues • Urgent emails • Attending an urgent meeting 3	**ELIMINATE** • Tasks that don't help to achieve your goals • Some subscription emails • Time-consuming nice-to-haves • Time wasting activities 4

Getoganizedgenie.com

This matrix helps you decide how much **time, energy, and resources** each decision deserves. For decisions with **high impact**, it's worth pausing and digging deeper, even if they don't feel urgent. Meanwhile, low-impact decisions shouldn't drain your energy.

How to Apply Triage in Real Life

Triage isn't just a tool for emergencies—it's a way to handle decisions every day. The first step is to pause and assess the potential impact, urgency, and complexity. Ask yourself:

- **What's the potential impact?** Will this decision have a major effect on you, others, or the broader picture?

- **How urgent is it?** Can this wait, or does it need an immediate answer?

- **How complex is it?** Do you need more information, or can a rule of thumb suffice?

Don't waste time on trivial or easily reversed choices and ensure you're giving the right attention to what matters most.

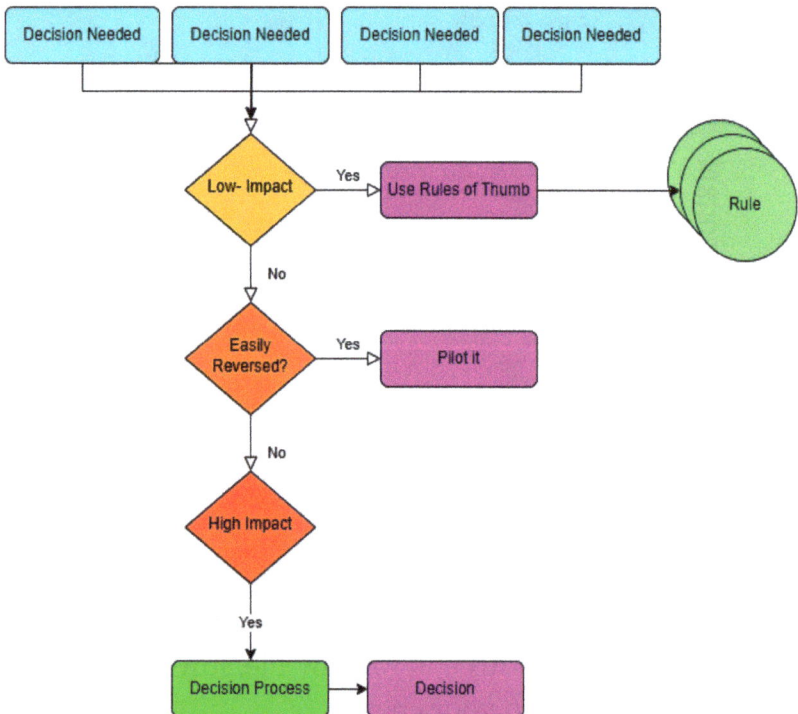

"Illustrations by Dall-E."

Whether it's deciding what to work on today or making a long-term life choice, triage helps you **focus your time, energy, and resources on what's important.**

Rules of Thumb for Low-Impact Decisions

Most decisions don't need deep analysis. For low-impact choices, simple rules of thumb can save time and energy while still leading to good outcomes. By committing to these rules ahead of time, you can **handle small decisions efficiently**. Here are a few examples:

Dessert once a week: A simple way to balance indulgence and health.

Follow the 2-minute rule: If a task takes less than 2 minutes, do it right away.

Buy quality, not quantity: Invest in fewer, better-quality items that last longer.

Sleep on big purchases: Wait 24 hours before making non-essential purchases to avoid impulse buys.

These rules make small decisions easier, freeing up your energy for the bigger, more complex choices.

Not every decision deserves the same time and attention. Sort decisions based on their potential impact, urgency, and complexity. Simple rules of thumb handle low-impact choices, leaving you more energy for the complex and emotionally charged decisions.

The key is knowing when to pause, assess, and decide how much effort each choice deserves. With practice, this approach can help you manage decisions wisely and avoid making choices in H.A.S.T.E.

That's My Perspective...

10-21-24
Decision Frameworks:

Mental Toolkits for Better Choices

"You've just finished triage on an urgent decision about the current cash crunch and a simultaneous huge opportunity. You must act decisively and within days, or the opportunity will be lost, and the business could be set back years—if you can stay liquid at all. You realize **this decision requires your full attention and the best tools you can muster**. What's next?"

"Bad decisions come from bad processes. A clear decision-making framework ensures you assess options effectively and act with confidence, without missing crucial details."

Better Decisions, Better Lives — Decision Education Foundation

"In this post, we'll explore three essential decision-making frameworks—the D.E.F. Decision Chain, the OODA Loop for rapid decisions, and the PDCA Cycle for continuous improvement. By the end, you'll have the mental tools to help you tackle high-stakes choices with confidence and precision."

The D.E.F. Decision Chain

"One powerful framework to improve decision quality is the **Decision Education Foundation (D.E.F.) Decision Chain**. Imagine this scenario: You're tasked with choosing between several business strategies. Each option seems viable, but you can't shake the feeling that you're missing something. This is where the Decision Chain comes in, guiding you through each critical step to ensure nothing gets overlooked."

The **Decision Chain** focuses on strengthening each link in the decision process, ensuring clarity and robustness at every step. It consists of five key components:

1. **Clarity of Values:** Understand what truly matters, why it's important, and how to ensure it aligns with those core values.

2. **Creative Alternatives:** Get creative. Generate a broad range of potential solutions, avoiding the trap of binary thinking. Mix and match ideas.

3. **Relevant Information:** Focus only on the data that impacts the decision and filter out noise.

4. **Sound Reasoning:** Apply logical and critical thinking to evaluate the alternatives objectively.

5. **Commitment to Follow-Through:** Once the decision is made, ensure there's a plan in place to execute it effectively.

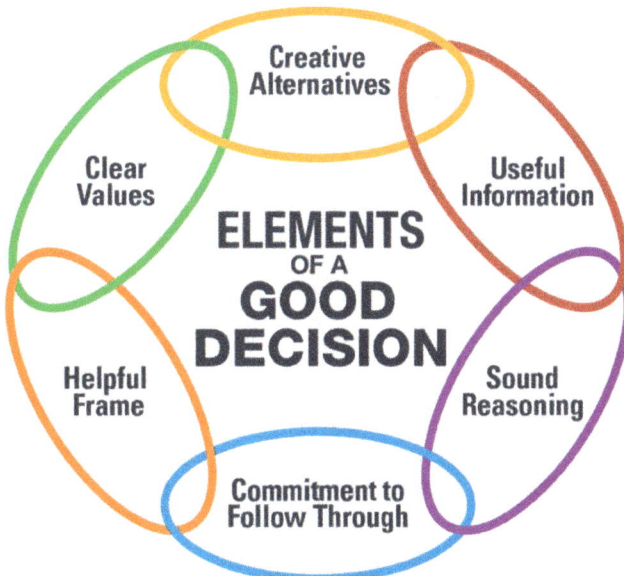

Creative Alternatives

Clear Values

Useful Information

ELEMENTS OF A GOOD DECISION

Helpful Frame

Sound Reasoning

Commitment to Follow Through

Decision Education Foundation

"The strength of the D.E.F. Decision Chain lies in its systematic approach to making sure that each component of the decision is thoughtfully addressed. If you miss just one link—such as failing to explore creative alternatives—you could end up with a limited and shortsighted choice."

The Decision Chain is only as strong as its weakest link. — Decision Education Foundation

The OODA Loop for Rapid Decision-Making

"In fast-moving environments, whether in business or personal life, decisions need to be made quickly and iteratively. The **OODA Loop**, developed by military strategist John Boyd, is a powerful framework for rapid decision-making. Created for fighter pilots in high-pressure situations, this model is now widely used in various fields, from business to emergency response."

The OODA Loop consists of four continuous steps:

1. Observe: Collect information from your surroundings. In a business context, this could mean gathering market data, customer feedback, or analyzing competitor behavior.

2. Orient: Make sense of the situation. Use your background, experience, and intuition to understand the context of your decision.

3. Decide: Based on the observations and orientation, choose a course of action. You may not have perfect information, but the key is to make a decision swiftly.

Act: Implement your decision immediately and observe the results, which feeds back into the cycle.

Observe… Orient… Decide… ACT!

The beauty of the OODA Loop lies in its adaptability. You are constantly gathering new data, reorienting, and adjusting your decisions. It's not about making one perfect decision but rather a series of smaller, more agile decisions that adjust to the situation as it evolves.

"Illustrations by Dall-E."

The Plan-Do-Check-Act (PDCA) Cycle

"When it comes to continuous improvement and refining processes, the **Plan-Do-Check-Act (PDCA) Cycle**, developed by W. Edwards Deming, is one of the most widely used frameworks. Originally designed for quality control in manufacturing, it has since been adopted across industries for iterative problem-solving."

The PDCA Cycle consists of four stages:

1. Plan: Identify a problem or goal and create a detailed plan to address it. This involves analyzing the current situation, defining objectives, and outlining the steps needed to achieve the desired outcome.

2. Do: Implement the plan on a small scale to test its effectiveness. This stage is about execution but in a controlled, low-risk manner to gather insights.

3. Check: Review the results of the test and compare them against the expected outcomes. Did the plan work as intended? If not, what went wrong? This stage is where lessons are learned, and adjustments are considered.

4. Act: Based on the insights gained in the Check phase, either scale the plan for broader implementation or go back through the cycle to make further refinements. The key here is to take what you've learned and act on it for continuous improvement.

"Illustrations by Dall-E."

Key Takeaways from Other Frameworks:

S.M.A.R.T. Goals Framework — "The S.M.A.R.T. framework ensures that decisions and goals are well-defined and actionable. By being **Specific**, **Measurable**, **Achievable**, **Relevant**, and **Time-bound**, it helps align decisions with clear outcomes and realistic timelines, making it ideal for goal-setting and project management."

Society of Decision Professionals (SDP) — Value-focused thinking shifts the focus from reactive decision-making to proactive decision-making. Clarifying values and objectives upfront ensures decisions align with long-term goals, making it highly effective for strategic decision-making.

The Cynefin Framework helps categorize problems into distinct domains: Simple, Complicated, Complex, and Chaotic. This allows you to adapt your approach depending on the nature of the problem, making it a valuable tool for navigating uncertain or volatile environments.

Six Thinking Hats

The Six Thinking Hats framework encourages decision-makers to see a problem from multiple perspectives. Using various thinking modes like logic, emotion, and creativity fosters balanced, innovative solutions, especially in groups that value diverse viewpoints.

The Weighted Decision Matrix or Pros and Cons framework, also called a "Rate and Weight" table, offers a structured way to compare multiple options by assigning weights to various factors. This method simplifies decision-making by breaking down complex choices into clear trade-offs, making it practical for everyday decisions.

Note: A great decision process leads to great results. Choosing the right framework for the situation is essential to success. Whether it's rapid action with the OODA Loop, continuous improvement with PDCA, or strategic alignment with Value-Focused Thinking, the framework shapes the decision and the outcome.

"The right decision-making framework can transform how you approach complex choices. "Each framework has strengths; the key is knowing when to use them."

Whether you're facing high-stakes business decisions or daily challenges, having a solid process ensures you don't overlook critical steps."

These frameworks equip you with the mental tools to tackle decisions with clarity, confidence, and precision. The next time you face a tough decision, take a moment to choose the framework that will guide you to the best outcome.

That's My Perspective...

10-28-24
The Art of Staying Cool:
How to Manage Stress
in High-Stakes Moments

Practical Techniques for Managing Stress and Making Confident Decisions

The alarm cuts through the quiet. Lights flash, and the firehouse comes to life. Firefighters jump from their beds, throw on their gear, and head straight to the truck. They're on the road within minutes, racing toward a four-alarm fire. Adrenaline is high—but they're calm, focused, and ready for whatever comes next.

"Illustrations by Dall-E."

Into the fire...

How is that possible? In moments where every second counts, how do they stay cool under pressure? It's more than just experience—it's about preparation, mindset, and controlling stress.

Knowing how to manage stress helps us make better decisions and stay in control when it counts. Whether it's mild anxiety or life-changing decisions, stress affects us all. From work to leadership to everyday life, staying calm under pressure is a key skill we all need.

Understanding our need for control is key to managing stress. When we feel uncertain, stress hormones trigger a "fight or flight" response. This can be helpful in real danger but overwhelming in everyday situations.

"Uncertainty is one of the most stressful things in life because we are hardwired to seek control over our environment," — Travis Bradberry, Emotional Intelligence 2.0.

Bringing the physical symptoms of stress under control is the first step. Techniques like breath control, muscle relaxation, and grounding exercises help calm the body and reset your response to stress. Once physical stress is managed, mental clarity follows.

Preparation, planning, and becoming antifragile are ways to regain control. After addressing physical stress, we can take further control through preparation and planning, anticipating challenges before they arise. By becoming more **antifragile**—able to grow stronger through stress—we limit uncertainty and create space for clear, calm decision-making.

Managing physical stress starts with controlling your breath. Under stress, our breathing becomes shallow and quick, amplifying the body's stress response. Slowing down and deepening your breath signals the body to relax. Box breathing (inhaling for 4 seconds, holding for 4, exhaling for 4, and holding again) can quickly calm the nervous system.

"Breathing is the quickest way to calm down in the moment and focus your mind." — psychiatrist Dr. Judith Orloff.

"Illustrations by Dall-E."

Physical relaxation techniques also break the cycle of stress. Progressive muscle relaxation, where you tense and then release muscle groups, helps reduce physical tension. Calming the body allows for mental clarity and better decision-making.

Preset rules simplify decision-making. Every firefighter knows one truth: **Stuff Happens**. On any call, things can go wrong, and fast. That's why they don't leave anything to chance. Before heading out, they run through **checklists** to make sure they don't forget essentials. But when they're on the scene, **preset rules and training** take over. They know exactly what to do because they've trained for it. These rules become automatic, freeing up their focus to handle whatever unexpected danger arises.

Preset rules free up mental energy for bigger decisions. By automating smaller, routine decisions, you conserve mental energy for more important choices. This helps prevent decision fatigue, where **too many choices lead to poor outcomes or burnout**.

Examples of some good preset rules:

- **Meetings:** Set a rule to only attend meetings with a clear agenda.

- **Snacking:** Drink a glass of water before reaching for a snack.

- **Money:** Automatically save a fixed percentage of every paycheck.

Developing your own preset rules is a form of preparation. The more you plan, the less room for uncertainty. Establishing rules or protocols ahead of time (because **Stuff Happens**) gives you structure when you need it most.

"**Be Prepared**" – Boy Scout Motto

Being organized is one of the best defenses against stress. When things are in order—whether it's your workspace, schedule, or thoughts—you can handle pressure more effectively. Disorganization adds to stress. Build systems and routines to help minimize the chaos that makes stressful situations worse.

Preparation creates a buffer against the unexpected and helps you to remain in control. Having a clear plan or knowing where to find information when needed gives you the confidence to stay calm. While organization won't eliminate stress, it ensures you're ready to manage it.

Organization helps you focus on what matters. When you're not overwhelmed by clutter, you have the mental space to focus on important decisions. Clear routines reduce distractions, allowing you to concentrate on critical tasks.

Thanks to my AI assistant Alfred and ChatGPT for help with developing this article.

That's My Perspective...

11-04-24
Make Any Decision Better:

Small Adjustments, Big Impact

1. The Key to Better Decisions.

What is the most useful thing you can do to improve any decision?

Answer: Get good feedback. Feedback from peers, experts, or real-world testing gives you the critical insights to refine your decision and adapt as new information comes in. Without it, you risk missing opportunities and failing to see potential pitfalls.

Feedback Provides Real-World Data. Piloting an option by running small experiments lets you gather concrete information about how your choices play out in the real world. This data surfaces blind spots that were invisible earlier.

Small adjustments, big changes in orbit.

Different Angles Unlock New Approaches. Get input from peers or experts to spot different angles, challenges, or solutions you might not have considered. Diverse feedback broadens your understanding and can offer ideas that strengthen your decision. Involve the stakeholders.

Feedback Enables Continuous Adjustment. Adapt in real-time by making small adjustments based on feedback. This keeps you on course and increases the likelihood of a successful outcome, avoiding the risk of waiting too long to correct.

"Illustrations by Dall-E."

2. Bring in Fresh Eyes

Get feedback by having stakeholders, peers, or experts review your plans. This helps reveal risks, alternative strategies, or ideas you might have overlooked, increasing your odds of a better result.

Get Real, Run Pilot Tests. Test likely options on a small scale, if possible. Pilot tests provide real-world data, allowing you to adjust before fully committing and avoid major pitfalls.

3. Think Through the Pitfalls

Ask **"What could make this the wrong choice?"** Considering potential failures helps you plan for them, strengthening your overall approach.

What Could Go Wrong? Identifying possible failure points allows you to build in contingencies and avoid problems before they become problems. By acknowledging risks upfront, you prepare for them instead of being blindsided.

What Has to Be True? For this option to remain the best choice, what conditions must continue to hold? Identifying these factors helps you monitor and adapt if they change.

The Role of Resilience: Adapting to New Information

"The best way to plan is to make very short plans, with a lot of optionality, and always keep many ways out." — Nassim Nicholas Taleb.

Build flexibility into your decisions so you can adapt when circumstances change. Resilient plans allow you to respond quickly without losing sight of your goals.

4. Set Your Course Markers

Define success and failure criteria. Establish what success looks like and, just as importantly, what signals failure. Clear criteria help you know when to stay the course, make adjustments, or pivot completely.

Set Triggers for Review and Change. Identify the key moments or conditions that signal it's time to reevaluate your decision. Setting triggers ensures you remain proactive rather than reactive.

5. Correct as You Go

Adjust like a guided missile. Just as a guided missile changes course to stay on target, make adjustments based on the feedback you receive. Flexibility is key to staying aligned with your goals.

"Illustrations by Dall-E."

Tiny adjustments

Make Mid-Course Corrections. Feedback allows you to adapt on the fly. Whether it's tweaking small details or making larger changes, staying flexible ensures your decision remains on track as circumstances evolve.

Be Open to Changing Your Mind. As you gather new information, it's smart to change your mind if the situation warrants it.

"When the facts change, I change my mind. What do you do, sir?"

- John Maynard Keynes

6. The Feedback Loop: Keep Improving

Use feedback to create a cycle of improvement. Continuous feedback—whether from real-world results or others' perspectives—ensures you're always refining your approach and learning from each decision.

Review, Reflect, Adjust, Repeat. Every decision improves with review, reflection, and adjustment. This loop creates a pathway for ongoing improvement, helping you make more informed and adaptable choices in the future.

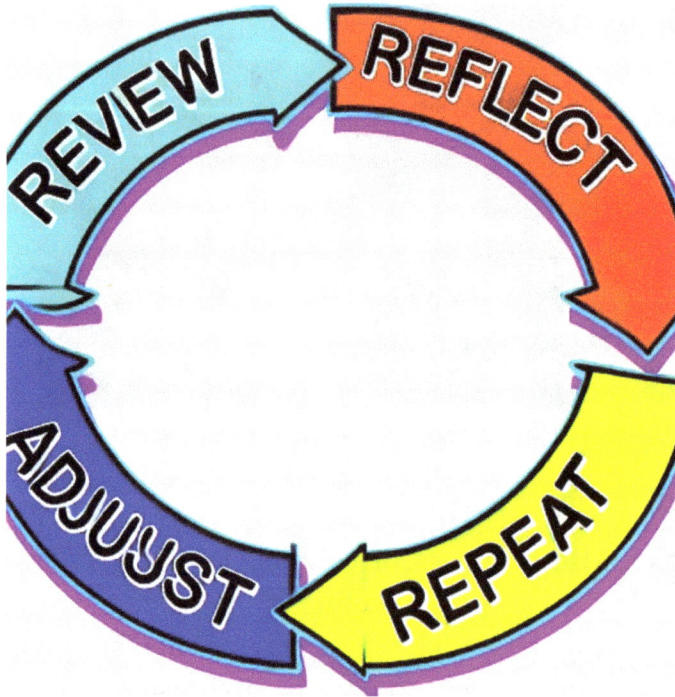

"Illustrations by Dall-E."

Review, Reflect, Adjust, Repeat

Feedback Drives Success. It lets you create a flexible approach that adapts to new information in every decision-making stage. This improves outcomes and reduces the risk of getting stuck in bad choices.

That's My Perspective...

11-11-24
Taking a Breather While 'My Perspective' Takes Off

On Publishing, Burnout, and the Next Chapter of This Journey

"My Perspective" is becoming a book.

I was deep in thought, hammering out a new blog post for the decision-making series, when my phone pinged, interrupting my flow. Muttering, I checked the notification, wondering who dared disturb the maestro. (Truthfully, I was blankly staring at the wall, struggling to finish the series' conclusion before I kick off the next one, "What to Do Once the Decision is Made.")

The call was from Bruce, my publisher. Years back, he helped me launch *A Concise Guide to Better Decisions*, so I assumed he wanted to catch up about that. But I was wrong.

After some friendly catching-up, Bruce cut to the chase: ***"I like following your blog. Let's make it a book."*** After a few femtoseconds of contemplation, I agreed. Bruce and I have worked together for years and have a strong friendship and high trust, so it only took a quick chat about file formats before I sent over all my posts from the last year. Now, he and his team are diving into edits, and there's more to come, so stay tuned!

But First, a Vacation.

Lately, I've noticed I'm nearing burnout. Working on complex AI prompts, mainly for educators, has been one of the most intellectually demanding projects I've taken on—fun, yes, but utterly draining. At the end of each day,

I feel like a wrung-out dishrag.

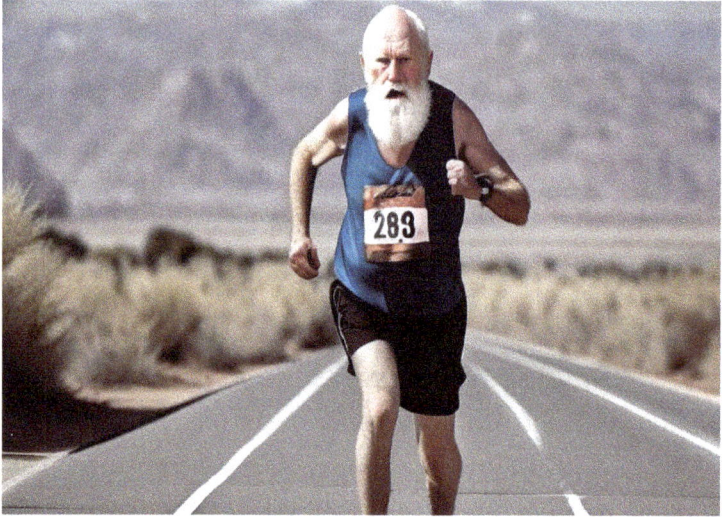

I Think I Can... I Think I Can...

The AI tools I've created cover everything from fact-checking to personal aides. In the process, I've earned a lot of *Experience*—the kind you gain by making every possible mistake...usually more than once. Some of these tools will be ready for you soon, so keep an eye out if you're into AI. But I digress.

The point? It's time for a breather. Officially "retired," I have no excuse. I'm taking responsibility for my own mental health, so I'm pausing this blog for a couple of weeks. When I'm back, I'll have fresh insights on turning decisions into the results you want—maybe even a prompt or two to share.

Stay tuned for updates on the book and AI tools—until then,

That's My Perspective...

12-02-24
Making It Real:

Make a Good Decision into a Great Result

Deciding what to do was the first step. **Making it happen? That's the key.** We've explored the 'Why' and the 'What' of good decision-making. Now it's time to dive into the 'How.'

"Without action, a decision is just a wish."

In this series, we'll break down **how to move from decision to action**—from setting up a strong foundation to building momentum and handling setbacks.

Meet the Team: The Essentials for Success

A strong team matters more than a perfect plan. A cohesive team can create a great plan; a fractured one will struggle, no matter how good the plan is. A solid team is like a well-oiled engine; without it, no matter how accurate your map is, you won't get far.

Whether you've built the team, inherited it, or joined as the newest member, here are some fundamental actions you can take to set the stage for success:

Psychological Safety and Belonging

Team members need to feel safe contributing and confident they belong. Without this, people hold back ideas, and collaboration suffers. A safe environment opens the door to innovation and teamwork.

Actions to Build Safety:

- **Encourage open discussion.** Let the team know all ideas are welcome and respectful disagreement is healthy.

- **Normalize mistakes.** Mistakes are valuable lessons. Remind the team that you've already paid the tuition—make sure to use the lesson.

- **Ask for feedback.** Regularly invite input on what's working and what isn't. It signals their opinions matter and helps you address issues early.

Practical Tip: Lead by example. Share your thoughts openly, especially when you're uncertain. Let the team know you value honesty and early communication, especially for bad news. This builds trust and ensures issues don't fester.

"Illustrations by Dall-E."

Trust...

Trust is the backbone of any successful team. When members trust each other, they collaborate openly and work toward shared goals. As the leader, your job is to ensure everyone understands and supports these goals.

Trust also ties directly to accountability. People need to believe that the team works toward shared objectives and that each member will deliver on their promises. Building trust requires clarity of purpose and consistent communication.

Practical Tip: Reinforce shared goals during regular check-ins. Ask how each task ties to the project's purpose. This keeps everyone aligned and reminds the team why their work matters.

Autonomy, Accountability, and Responsibility

People do their best work when they're trusted to make their own decisions. Giving team members ownership of their tasks keeps things moving smoothly. But autonomy needs accountability—team members must feel responsible for their work and the team. Responsibility ties it all together: it's the commitment to follow through and support each other.

Build an environment where autonomy, accountability, and responsibility reinforce one another. When people are trusted to decide, committed to their work and teammates, and expected to deliver on promises, the team builds momentum naturally.

Clarity, Goals, and Commander's Intent

"Illustrations by Dall-E."

What's our goal?

When goals are unclear, projects drift. If the team doesn't know what they're aiming for, they'll move in different directions. Your job as a leader is to define the goal clearly and explain the reasoning behind it. This is where **Commander's Intent** becomes invaluable.

Commander's Intent goes beyond the "what" by explaining:

- **Why** the goal matters.

- **How** the decision was made.

- **Limits** or "rules of engagement" (e.g., budget constraints, no customer outages).

When the team understands not just the goal but the reasoning and boundaries, they can make better decisions independently while staying aligned with the project's purpose.

Practical Tip: Along with your goals, create a "Not-to-Do" list. This helps the team focus and prevents scope creep—when projects grow beyond their intended scope. Regularly revisit both the goals and boundaries to ensure the team stays on track.

"If you don't know where you're going, you'll end up someplace else." – Yogi Berra

Starting a project with the right foundation is half the battle. From building a strong team to setting clear goals, each step brings you closer to turning decisions into results. Simplicity, clarity, and alignment with stakeholder needs (those directly affected by the project) are key to ensuring the project doesn't just get done—but gets done well.

With the groundwork set, the next step is about keeping the momentum going. In our next post, we'll focus on building a daily routine that keeps the team moving forward, even when challenges arise.

Special thanks to Al, my ChatGPT AI assistant, for helping me research and build this post. By the way, the illustrations are Al's too.

That's My Perspective...

12-09-24
Staying Focused:

Keeping the Team on Track

Setting up a strong foundation is just the beginning. The real test lies in maintaining focus when priorities shift, challenges emerge, and new demands arise. Keeping the team grounded and adaptable ensures the project stays on track.

"Illustrations by Dall-E."

Focus is what matters...

Connect to Reality

Stakeholders can change during the project. Stakeholders— people affected by your project, such as end users, customers, executives reviewing reports, or budget managers tracking expenses—often have evolving needs. Business priorities shift, and external factors can crash into even the best-laid plans. Staying grounded means keeping the team connected to what truly matters through regular feedback loops and adaptable planning.

"The successful warrior is the average man, with laser-like focus." — Bruce Lee

Understand the critical players. Stakeholders are individuals or groups with a vested interest in the outcome. Knowing who they are and what they need is key to keeping the project relevant.

Adapt and check in regularly. Flexibility is essential. Regular check-ins with stakeholders ensure the team can adjust to changes and deliver real value.

Short-term plans keep things flexible. Breaking big goals into manageable steps helps the team stay responsive and effective as conditions change.

Practical Tip: Establish regular feedback sessions with stakeholders. Open, honest communication ensures the team stays on track and aligned with the project's purpose.

Avoid Gold-Plating

Simplicity is powerful. The urge to over-deliver often leads to adding unnecessary features—a practice known as gold-plating. While it might seem impressive, gold-plating wastes time, resources, and can even introduce risks.

It's tempting to add every good thing you can think of, but every new feature also brings extra cost, time, and complexity. The result? Delays and distractions from the project's core goals.

Focus on the essentials. Stick to stakeholder-defined "must-haves" and leave "nice-to-haves" for later.

Practical Tip: Start with a "minimum viable product" (MVP) that delivers the essential functionality. Use feedback from the MVP to guide additional features based on real-world insights.

Watch Out for Hidden Incentives

Hidden incentives quietly shift focus. Stakeholders or team members may push for features that serve personal preferences or prestige rather than project goals. These can quietly derail progress.

Bring every request back to the goal. Evaluate how each request aligns with project objectives and call out anything that doesn't. Ask, "How does this serve the project's goals?" This keeps the team focused on delivering meaningful results.

Preparation and Planning Essentials

Big tasks can feel overwhelming. Breaking them into smaller, phased steps makes them manageable. This reduces stress and keeps progress steady.

"A goal without a plan is just a wish." — Antoine de Saint-Exupéry.

Focus on the "monkey," not the "pedestal." Astro Teller's "Monkey and Pedestal" analogy explains this perfectly: If your goal is to teach a monkey to juggle while standing on a pedestal, start with the juggling. The pedestal is easy—if you can't teach the monkey to juggle, there's no point in building it. Tackle the hardest, most critical tasks first. (Source).

Understand the critical path. The critical path is the sequence of tasks that determines the project's completion time. These are often the riskiest and most complex steps. Addressing them early prevents bottlenecks and smooths the way for the rest of the project.

Practical Tip 1: Identify the toughest challenges (the "monkeys") and tackle them first. This ensures the critical path is addressed early, making progress smoother.

"Illustrations by Dall-E."

Practical Tip 2: Assign clear ownership for each task. Even if a group works on it, one person should be responsible for ensuring it's done right. Ownership prevents confusion and keeps tasks on track.

Momentum Matters

Staying focused is an ongoing effort. A strong start doesn't guarantee success. Challenges and distractions will arise. The key is maintaining momentum by solving problems early, focusing on essentials, and aligning the team.

Momentum comes from consistency, not just progress. Small wins—like solving a tricky problem or hitting a key milestone—build morale and keep the team moving forward.

Practical Tip: Celebrate progress often. Call out both team and individual efforts to keep spirits high. Reflect regularly on what's working and what's next—it keeps everyone motivated and on track.

Starting strong is important, but **staying focused makes a project succeed**. Keeping the team grounded, tackling the hardest problems first, and maintaining steady momentum allow you to turn plans into real results—even when challenges arise.

Special thanks to Al, my ChatGPT AI assistant, for helping me research and build this post.

That's My Perspective...

12-16-24
A Small Gift

It's that time of year

As we get close to the end of the year, I thought it would be appropriate for me to share a small gift with you. I have a template I use for my personal journal that I hope you might find useful. I've created a "Journal" section in my OneNote notebook, so it's available whenever I want to capture something. If you're into journaling, which I highly encourage, here's a cure for the blank page and a thought-starter.

"Illustrations by Dall-E."

Open loops: What tasks or thoughts do I need to capture and process?

Gratitude: What are three things I'm grateful for today?

Successes: What went well, and what strengths contributed to that success?

Lessons: What didn't go well? What can I learn or adjust moving forward?

Priority: What is the top priority I want to focus on tomorrow?

Anticipate Challenges: What obstacles might arise, and how can I prepare to handle them effectively?

Next Actions: What is the first step I need to take for my top priority?

One Improvement: What's one small thing I can do differently tomorrow to make it even better than today?

Thoughts:

That's My Perspective...

12-23-24
What Thriving Really Takes

Mastering the skills school left out.

Schools teach us facts but don't teach us how to thrive. You probably learned how to solve for x, memorize historical dates, and write essays. But did anyone teach you how to decide where to live, how to pick your friends, or what kind of work would give your life meaning? Did you learn how to keep your cool when everything went wrong—or how to support a friend through a tragic loss? **For all the good our schools do, they don't prepare us to thrive.**

The Problem

Schools don't give us all the tools we need to thrive. Math, history, and writing are useful, but they don't prepare us for life's hardest decisions: where to live, what career to pursue, or who to build a life with. **They don't teach us how to handle failure, support loved ones through loss, or manage a financial crisis.** But you can fill in these gaps.

Thriving isn't automatic—it takes intentional effort and the right tools.

Why It Matters

Thriving requires skills that life demands. Decision-making, emotional intelligence, and resilience aren't extras—they're essential. Sooner or later, everyone steps into deep waters. Usually, you're left to sink or swim.

"Illustrations by Dall-E."

These skills shape the quality of your life. They help you evaluate tough choices, recover from setbacks, and build meaningful connections. They let you handle life's highs and lows with class.

But we're rarely taught these skills. Most of us stumble along through trial and error, learning lessons the hard way. It doesn't have to be this way.

What Thriving Takes

Thriving isn't about luck—it's about mastering the right tools. These tools aren't mysteries; they're skills anyone can learn with effort and practice. Decision-making frameworks help us weigh choices effectively, as Annie Duke explores in *Thinking in Bets*. Emotional intelligence lets us navigate relationships and challenges with grace, as described in *Emotional Intelligence 2.0*. As Rick Hanson teaches in **Resilient**, resilience keeps us moving forward when life gets hard.

These tools can help you handle reality. Imagine making tough decisions with confidence, staying calm in a crisis, or building trust in every relationship. These skills don't just make life easier—they make it richer and more fulfilling.

And it's never too late to learn them. Whether you're trying to dig out of a hole or looking to improve your already awesome existence, you can master the skills that build a thriving life.

My mission in life is not merely to survive, but to thrive; and to do so with some passion, some compassion, some humor, and some style."

— Maya Angelou

You know where you're hurting most. I wish I'd known about these as I've dealt with my own problems. **Thriving isn't a goal—it's a skillset we can all develop.**

Additional Resources:

1. Decision-Making

- "Thinking in Bets" by Annie Duke
  ISBN: 978-0735216358

- "Decisive: How to Make Better Choices in Life and Work" by Chip Heath and Dan Heath
  ISBN: 978-0307956392

- "Clear Thinking: Turning Ordinary Moments into Extraordinary Results" by Shane Parrish
  ISBN: 978-0593489481

- **Alliance for Decision Education** Website: https://alliancefordecisioneducation.org/

- **Decision Education Foundation** Website: http://www.decisioneducation.org/

2. Emotional Intelligence

- "Emotional Intelligence 2.0" by Travis Bradberry and Jean Greaves
  ISBN: 978-0974320625

- "Nonviolent Communication: A Language of Life" by Marshall B. Rosenberg
  ISBN: 978-1892005281

- "Dare to Lead" by Brené Brown
  ISBN: 978-0399592522

3. Resilience

- "Resilient: How to Grow an Unshakable Core of Calm, Strength, and Happiness" by Rick Hanson
  ISBN: 978-0451498847

- "The Resilience Factor" by Karen Reivich and Andrew Shatté
  ISBN: 978-0767911917

- "Grit: The Power of Passion and Perseverance" by Angela Duckworth
  ISBN: 978-1501111112

4. Financial Literacy

- "Rich Dad Poor Dad" by Robert T. Kiyosaki
  ISBN: 978-1612680194

- "The Total Money Makeover" by Dave Ramsey
  ISBN: 978-1595555274

- "Your Money or Your Life" by Vicki Robin and Joe Dominguez
  ISBN: 978-0143115762

5. Critical Thinking

- "Thinking, Fast and Slow" by Daniel Kahneman
  ISBN: 978-0374533557

- "The Demon-Haunted World: Science as a Candle in the Dark" by Carl Sagan
  ISBN: 978-0345409461

- **"Critical Thinking: A Beginner's Guide" by Sharon M. Kaye**
  ISBN: 978-1851686541

6. Communication

- "Crucial Conversations: Tools for Talking When Stakes Are High" by Kerry Patterson et al.
  ISBN: 978-0071771320

- "How to Win Friends and Influence People" by Dale Carnegie
  ISBN: 978-0671027032

- **"Difficult Conversations: How to Discuss What Matters Most" by Douglas Stone, Bruce Patton, and Sheila Heen**
  ISBN: 978-0143118442

7. Collaboration

- "Team of Teams: New Rules of Engagement for a Complex World" by General Stanley McChrystal
  ISBN: 978-1591847489

- "The Five Dysfunctions of a Team" by Patrick Lencioni
  ISBN: 978-0787960759

- **"Collaborative Intelligence: Thinking with People Who Think Differently" by Dawna Markova and Angie McArthur**
  ISBN: 978-0812994902

8. Self-Regulation

- "The Power of Habit: Why We Do What We Do in Life and Business" by Charles Duhigg
  ISBN: 978-0812981605

- **"Atomic Habits: An Easy & Proven Way to Build Good Habits & Break Bad Ones" by James Clear**
  ISBN: 978-0735211292

9. Purpose and Values Identification

- "Man's Search for Meaning" by Viktor E. Frankl
  ISBN: 978-0807014271

- "The Seven Habits of Highly Effective People" by Stephen R. Covey
  ISBN: 978-0743269513

- **"Start with Why: How Great Leaders Inspire Everyone to Take Action" by Simon Sinek**
  ISBN: 978-1591846444

10. Plays Well with Others

- "Plays Well with Others: The Surprising Science Behind Why Everything You Know About Relationships is (Mostly) Wrong" by Eric Barker
  ISBN: 978-0062884005

Notes:

Notes:

www.ingramcontent.com/pod-product-compliance
Lightning Source LLC
Chambersburg PA
CBHW051715020426
42333CB00014B/991